IMAGES
of America

INDIO'S DATE FESTIVAL

On the Cover: Queen Scheherazade contestants pose for publicity photographs behind the Old Baghdad Stage around 1948. Since 1947, publicity images often feature young women in Arabian-inspired costumes as well as other iconic features of the date festival such as the Arabian-inspired architecture, camels, or even dates. (Courtesy of the Coachella Valley History Museum.)

IMAGES
of America

INDIO'S DATE FESTIVAL

Sarah Seekatz

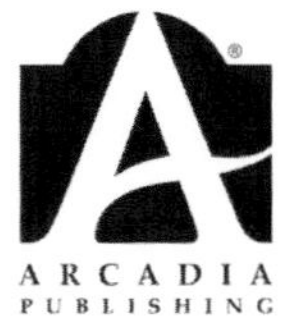

Copyright © 2016 by Sarah Seekatz
ISBN 978-1-4671-3425-5

Published by Arcadia Publishing
Charleston, South Carolina

Printed in the United States of America

Library of Congress Control Number: 2015936506

For all general information, please contact Arcadia Publishing:
Telephone 843-853-2070
Fax 843-853-0044
E-mail sales@arcadiapublishing.com
For customer service and orders:
Toll-Free 1-888-313-2665

Visit us on the Internet at www.arcadiapublishing.com

To Charlotte

Contents

Acknowledgments

Like dates, this book was cultivated and nourished by many people along the way. My thanks go first to the participants, staff, and volunteers who have supported the date festival over the years and especially the families who allowed me to publish their treasured images in this book. I am particularly grateful to those who assisted me at the Coachella Valley History Museum, where many of these photographs are preserved. Erica Ward, Janice Woodside, and especially Robert Tyler generously guided my exploration of Indio's history there. The long-term volunteers such as Connie Cowan, Patricia Laflin, and Louise Neeley influenced this work as well through their histories of the date industry and their organization of the date archives.

Completed just after my graduation, this work was heavily shaped by research conducted for my dissertation. As such, I must extend my thanks to those who supported that project, particularly Molly McGarry, Catherine Gudis, and Vicki Ruiz, as well as the archivists throughout California who preserve, often from afar, the Coachella Valley's history. Caitrin Cunningham and the staff at Arcadia deserve credit for helping me format this work, while my eternal thanks go to Beverly Davis for her outstanding editing skills and friendship.

My thanks extend to my incredibly supportive family. I am particularly appreciative of my parents, Patrick and Margo McCormick, who encouraged this work and to my sister, Amanda, who gave me the gift of quiet time. The support of my aunt and resident librarian Ruth McCormick and the thoughtfulness of my in-laws Peter and Suzanne Seekatz helped this work as well. Finding a photograph of my grandmother, Patricia Salcido Ligman, dressed in her pageant outfit, remains a research highlight for me; I thank her for cheering me on from afar. As with everything, I could not have written this book without the love and kindness of my husband, Scott. And my dear little Charlotte, thank you for keeping me company as I wrote this book. Despite all the time invested between these pages, you and your dad are still my very favorite dates.

Unless otherwise noted, all images appear courtesy of the Coachella Valley History Museum.

Introduction

A visit to the Riverside County Fair might at first seem like any other American festival. Rickety roller coasters entertain kids, grandmas submit homemade pies and jams for judging, and families enjoy the petting zoo. But a trip to this county fair reveals a much more exotic setting. The fair's princesses pose for photographs, like they do at so many other fairs across the nation, but the royalty here look like they have been airlifted in from Disney's *Aladdin*. They wear midriff-baring "Middle Eastern" garb and answer to the names Queen Scheherazade and Princess Jasmine. Visitors await the pageant with anticipation, catching the performances of loved ones on stage. This pageant, however, sends visitors back hundreds of years and thousands of miles to the Middle East of *One Thousand and One Arabian Nights*, transporting them with suggestive costumes and a stage designed to look like old Baghdad. Fairgoers cheer at pig races and then take in the camel races. This is not just the Riverside County Fair; it is also the National Date Festival. For the community of Indio, the date is more than a cash crop; it is an excuse to play Arab for two weeks every year.

Dates, though an important product in the Coachella Valley, are not native to California. The US Department of Agriculture first imported date palms from the Middle East and Northern Africa for large-scale cultivation around the turn of the 20th century. Government scientists worked with locals to encourage investment in date groves and harnessed their knowledge of agriculture to improve fruit output. By the time the small offshoots matured enough to produce a commercial crop, Indio and its surrounding communities had already begun to celebrate the date. The city held the first Festival of Dates in 1921, hosting fairs periodically until it became an annual event in 1947.

For the most part, these celebrations went hand-in-hand with a borrowed notion of the so-called Orient. Influenced by religious and popular culture, Americans long romanticized the Middle East, seeing the region as both a lush, rich, exotic, and sexualized space and a place of poverty, disease, and filth. This Orientalism allowed the West to define itself by what it was not, a civilized modern society compared to visions of a backward and ancient Middle East. Such views defined popular culture even while Americans fell in love with all things Arabian. Films like *The Sheik* and *Cleopatra* reflected the theme, while high fashion embraced harem pants. Consumer culture turned to the Orient as well, utilizing the Middle East's romance to sell everything from soap to soup. Travel literature paired with new forms of photography (the stereopticon, magic lantern slides, and eventually reproduction technology for newspapers and magazines) to sell romance to the armchair traveler. Of course, people had dreamed of the Middle East long before, with the Bible and *One Thousand and One Arabian Nights* daily reading for many, shaping their views of the region as early as the American Revolution. The 1922 discovery of King Tutankhamun's tomb added a layer of Egyptomania to America's love affair. Movie palaces, including Indio's Egyptian Theater, embraced Egyptian theming while the flappers' bobbed haircuts and fashion styling reflected ancient Egyptian depictions.

To set their communities apart as tourist destinations, early Coachella Valley boosters turned to this Orientalism to remarket the area. The similar desert climate and landscape, paired with the recently relocated date palms, proved easy inspiration for this transition. First, locals renamed towns. Walters became Mecca, Durbrow became Arabia, and towns like Oasis, Thermal, and Edom also referenced the Middle East. Real estate developments were also proposed: the Walled Oasis of Biskra, though never completed, was set to have North African architecture and Moorish-style shops. Potential investors arrived in Indio by train and then rode camels to the site, guided by handlers in sheik costumes. Date shops also embraced Arabian-inspired design. Sniff's date gardens not only mirrored this style but even displayed a Bedouin tent in their gardens. Indio housed the Garden of Eden, a pyramid-shaped date shop whose name, like so many others, referenced the Orient as well. Traces of this Middle Eastern influence can be seen even today in Coachella's Cairo Street, Coachella Valley High School's Arab mascot, and Indio's Arabian Gardens Trailer Park.

Thus it is little wonder that the Arabian theme took off at the date festival, especially as fair organizers tried to differentiate themselves from other Southern California fairs in the late 1940s and early 1950s. The fair maintained a Western theme in the years prior to World War II, a theme that several other celebrations utilized around the Southland. As the only Arabian-themed festival, organizers hoped to draw in more publicity and thus tourism to the Eastern Coachella Valley. Boosters suggested it was also a better fit to the region, given the similar climate and landscapes, as well as the unique agricultural crop of the date. Many of the Arabian mainstays, including the Moorish architecture, camel races, Queen Scheherazade beauty pageant, and *Arabian Nights* musical pageant, arrived under the direction of Robert Fullenwider, fair manager, though countless others supported and organized these events. Everyday residents played their part as well, particularly through the donning of "Arabian" garb during fair time. Switching from cowboy attire to harem pants and fezzes, locals felt as though wearing costumes was a way to boost their city. They believed that, if visitors saw the town full of sultans, caliphs, and queens, they would remember and return to the fair yearly. Perhaps the costumes might inspire some to relocate to a town with residents that not only had the money to dress up but the community spirit to do so as well. Merchants in particular embraced the costume as a fair-time uniform for their employees, hoping to draw in additional business as well as boost the community.

Though locals sought to celebrate the Orient at the fair, the appropriation of Middle Eastern imagery had a downside as well. Often, the Arabian theme reinforced the larger stereotypes of Orientalism. The pageant, which followed the storylines of *One Thousand and One Arabian Nights*, depicted murderous caliphs, evil genies, a thieving population, and slaves. Although the stories of the *Arabian Nights* were set in the medieval time period, many visitors translated these plot lines to the current Arab world. This, by their imagination, rendered its population as dangerous, violent, and backward, never evolving with contemporary society.

Popular culture already sexualized the Middle East. *The Sheik* was sex-crazed in 1921. Elvis Presley's 1965 *Harum Scarum* featured the ubiquitous Arab seductress among its scantily clad harem girls. Likewise, the harem girl costumes at the fair reinforced stereotypes of the Orient, particularly ones aimed at Arab women. The poses of queen contestants, often around a male fair employee, recreated the harem. In doing so, photographic representations, reproduced in newspapers around the nation, implied that women of the Middle East were readily sexually available to the Western male. Sex sells, and at the date festival, high crowns and low necklines meant that women's bodies came to market the fair and the crop to the larger world. When publicity photographs made their way into the media, they nearly always featured young women in the harem outfit, beckoning visitors to explore the warm desert, fun festival, and delicious dates they represented.

Fair organizers seldom claimed to recreate an authentic version of the Middle East. Time and again they declared their costumes and fair an event influenced by Hollywood as well. However, since Indio had ties to the region, especially through the exchange of scientific knowledge on the date palm, they were seen as an authority on all things Arabian. In a time when few Americans claimed Middle Eastern ancestry, Indio was the closest most people got to experiencing or learning about the area. As popular culture increasingly rendered it a dangerous place, where terrorists and greedy oil barons abound, the date festival stood out in its attempts to celebrate the birthplace of the date.

Before immigration from the Middle East and Northern Africa increased after 1965, few people from that part of the world visited the festival or the Coachella Valley. When they did, it often became a newsworthy event. Newspapers reported that when international royalty visited the date groves or the fairgrounds, they proclaimed the beauty of the California desert and appreciated the references to their homelands. But these stories were always filtered through the local media; we simply do not know how these Middle Easterners actually perceived the Orientalism at the fair. Few people of Middle Eastern ancestry called the Coachella Valley home. The 1930 census of Indio, for example, listed just 20 people who were Syrian or Syrian American, out of a population of 3,486. There were no other people of Middle Eastern descent. These two families, the Peters and Abrams, held the privilege and burden of representing the Middle East to their fellow Coachella Valley residents. With less than half of one percent of the population claiming descent from Arab nations, this underrepresentation continues in Coachella Valley communities even today.

Over the past 40 years, the emphasis on the Middle East has lessened, in part because of the changing relationship American popular and political culture holds with the Middle East and Northern Africa. Nightly news broadcasts have reported on the oil embargo, Iranian hostage crisis, Persian Gulf War, and military involvements since 9/11, bringing a new vision of the Middle East into American homes. The date festival never abandoned its Arabian theme, but, as the 20th century progressed, the community grew less inclined to embrace Middle East–inspired costume and pageantry. A changing populace also contributed. As the Eastern Coachella Valley's population boomed, fewer of its residents had direct ties to the date industry. The fair's focus shifted towards headliner entertainment and standard county-fair treats like carnival rides and corn dogs. Held every year in February, the Riverside County Fair and National Date Festival remains one of Southern California's most unique fairs, celebrating a crop as ancient as civilization itself.

One

Early Festivals
From Wise Men to Whiskerinos

Agricultural festivals across the nation have long celebrated local specialties; in the 1910s, the Coachella Valley held festivals for annual harvests. Some of these events specifically championed the growing date industry. These local fairs brought together the community, but area boosters also displayed their dates at regional celebrations such as the Riverside Fair. Indeed, the date industry represented the county at the 1915 Panama Pacific International Exposition (world's fair) in San Francisco. Local grower W.L. Paul and his family encouraged visitors to travel to the date groves, displayed photographs of date fields, directed people to large date palms on display, and even distributed samples of the valley's delicious dates. The tradition of exhibiting the date industry at larger state events continued. Date shows exhibited the fruit at the Southern California Fair (and National Date Show) in Riverside in the 1920s and early 1930s, while special date exhibits reached the California State Fair as well. The California Date Exhibit opened to an audience of two million people at the Golden Gate International Exposition in 1939 and 1940.

The first festival to draw more than a local crowd to the Coachella Valley was the widely publicized 1921 International Festival of Dates, which featured an Arabian theme. Boosters hoped to continue the event annually but were only able to host a few more regional fairs in the 1920s. Though festivals around the state closed or merged together during the Great Depression, the Riverside County Fair came to Indio in 1938, brought there with the hard work of the Indio Civic Club, which desired to combine the event with their regional date celebrations and stampede rodeos. Until World War II stopped all fairs in the state in 1942, the event maintained a Western theme. Many of these early festivals were set in what is now known as Miles Park, not far from downtown Indio.

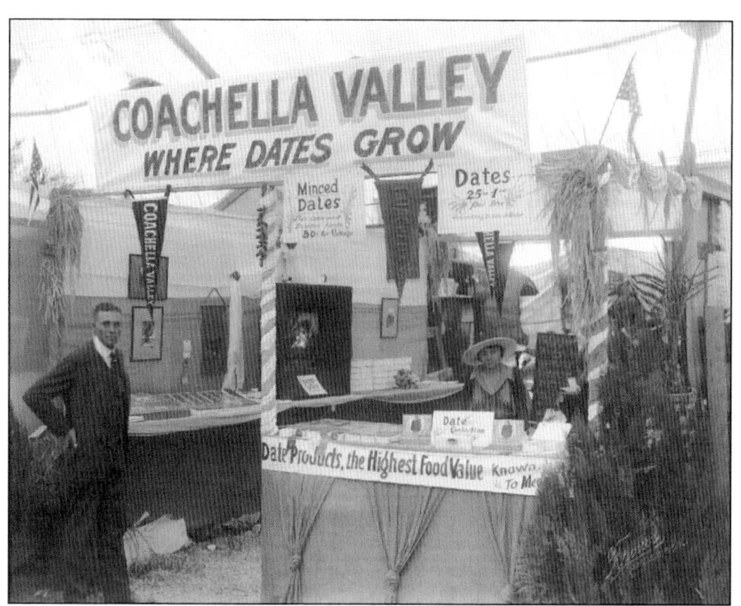

Boosters in the valley saw regional fairs as an opportunity to encourage Californians to relocate and invest in dates. Here, W.L. Paul and his daughter Nina Paul Shumway occupy a booth at the 1913 Riverside Fair. Booths like this drew attention because the fruit was so exotic. The Pauls sold dates and date confections, thereby combining their community investment with their business ventures.

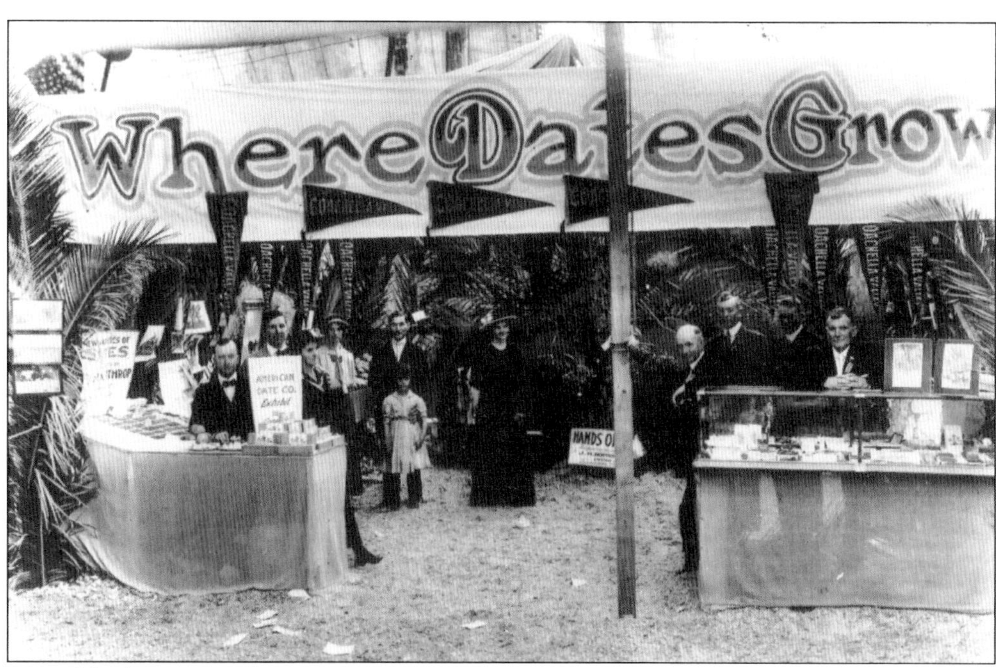

One of the few images of the 1921 Festival of Dates, this photograph includes the American Date Company Exhibit. The company built the first packinghouse in the Coachella Valley in 1912. The exhibit included several date palms with the fruit still on, an educational display provided by grower J. Northrup, one of the earliest commercial date growers in Indio.

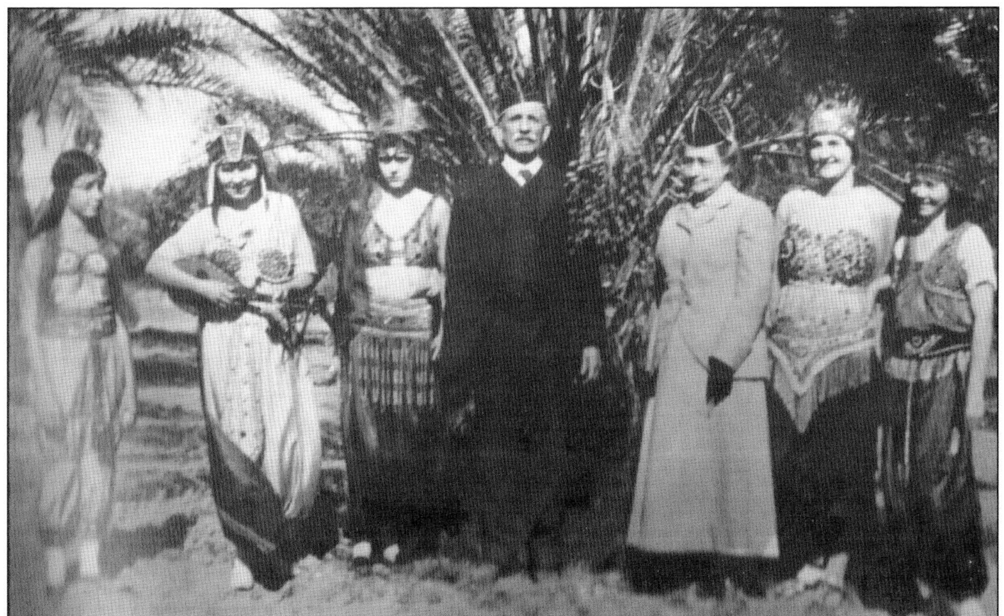

To embrace the Arabian theme, attendants at the entrance gate wore Arab garb while some local women donned harem outfits. The harem-girl fashion burst onto the French scene in the 1910s with designs by Paul Poiret. It was also popularized by the "hootchie coochie" (belly) dance prominent at world's fairs. By 1921, the United States was well versed in cross-cultural dressing, displayed here by local women's harem-girl costumes.

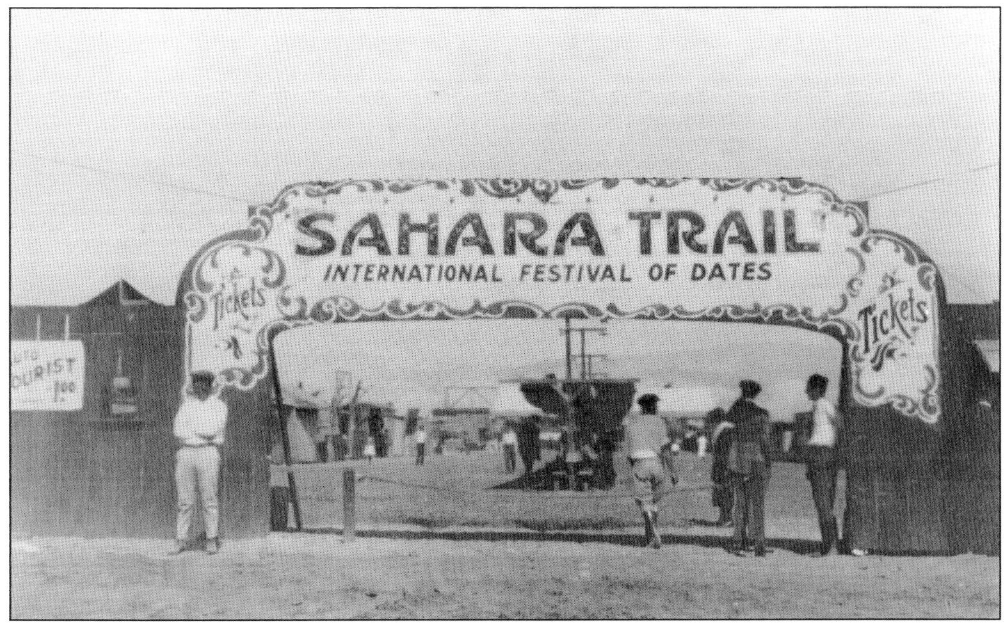

The 1921 festival touted itself as the Sahara Trail, just one of references to the Middle East and Northern Africa. The event lured in travelers from around Southern California, despite a six-hour trip from Los Angeles. The county invested in road improvements for the event while fair organizers made use of the improvements by organizing a caravan from Indio to Los Angeles, stopping at each city to pass out flyers. (Courtesy of the Coachella Valley Water District.)

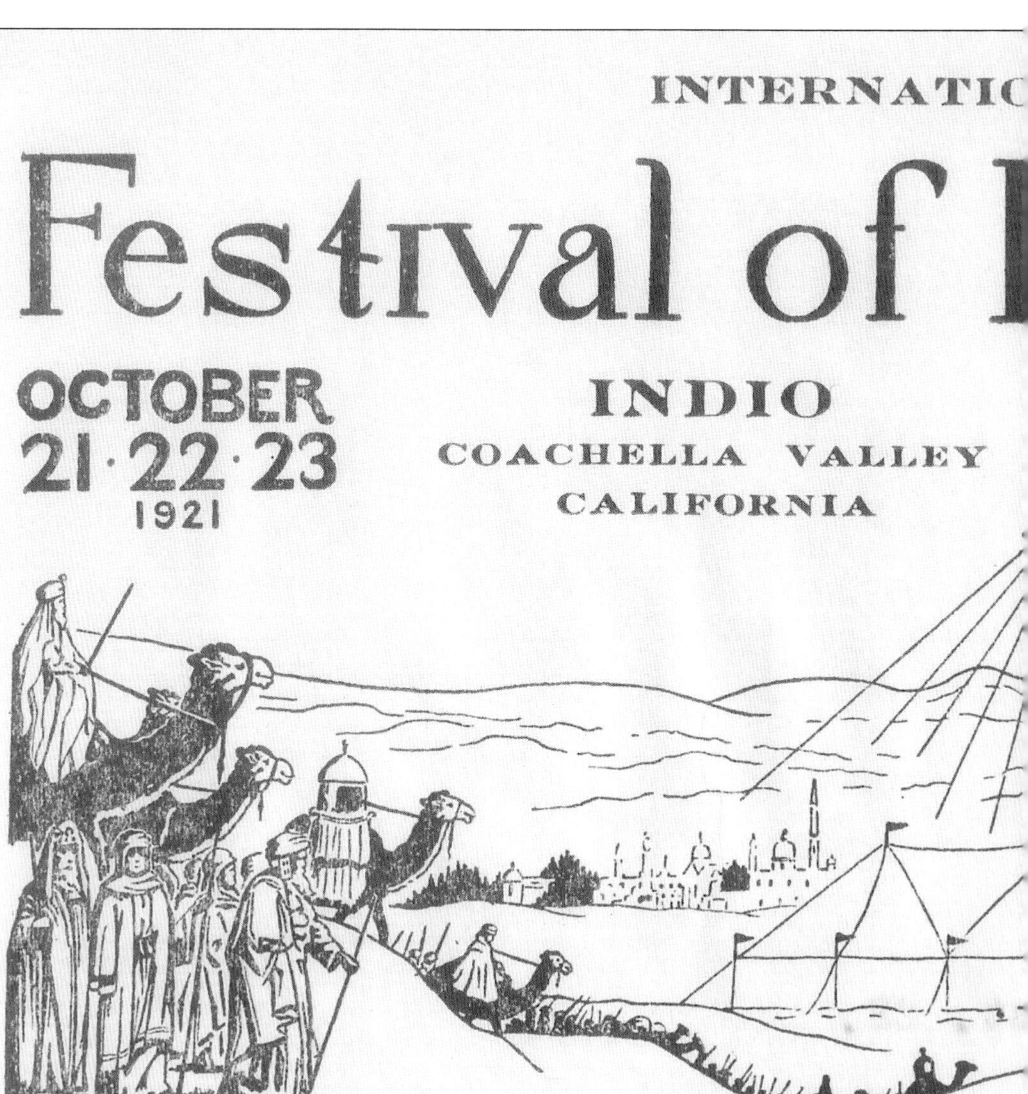

This advertisement for the first Festival of Dates positioned the blossoming date industry as the valley's literal messiah with its overt references to the nativity narrative. The flyer invokes the Star of Bethlehem, while lines and lines of wise men journey toward it. Implying that wise men would not only visit the fair but invest as well, the biblical imagery connected visitors to one of their more familiar understandings of the Middle East. While the international designation

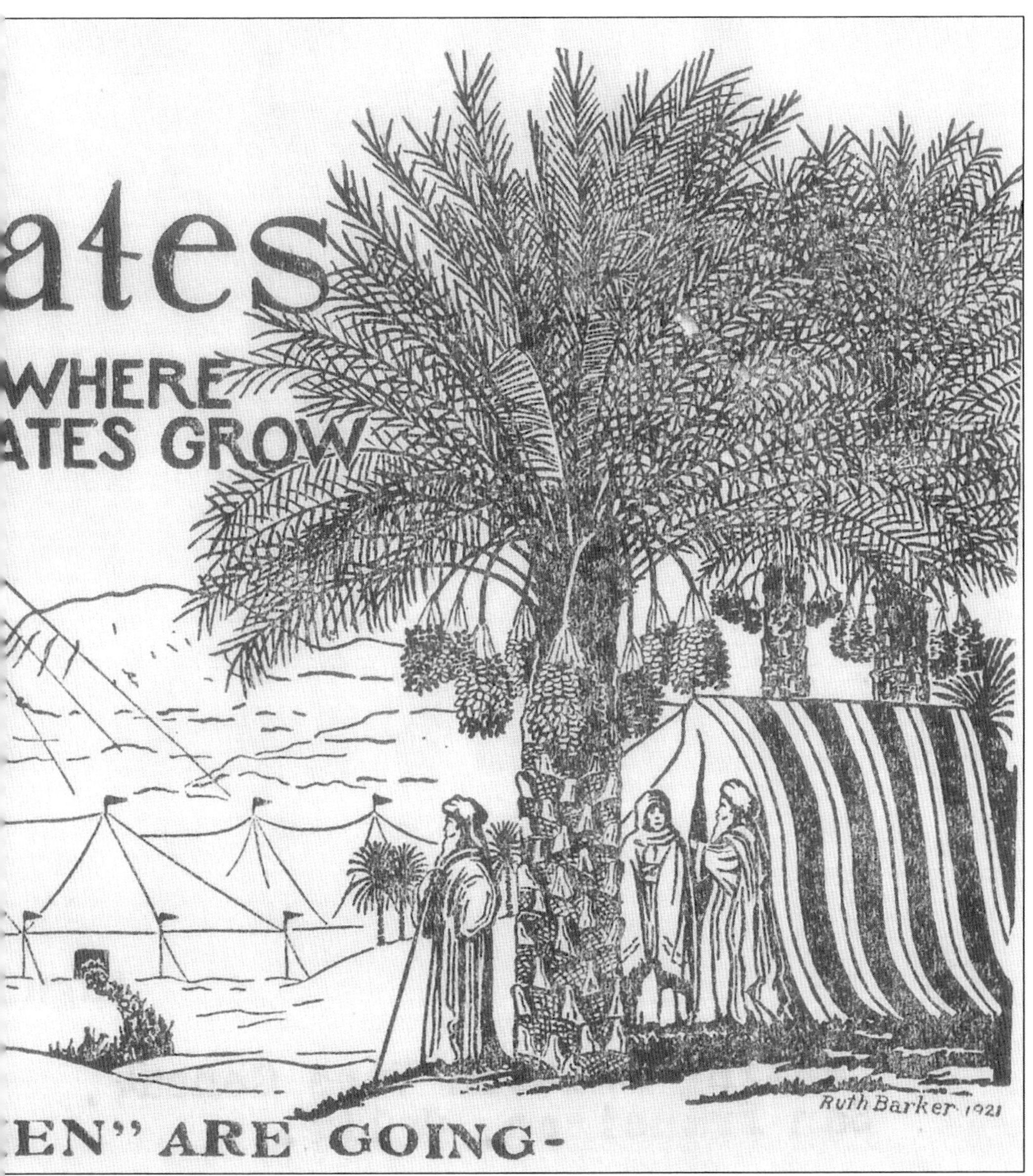

invoked the Orient, in reality just two nations sent low-level representatives: Mexico, whose Baja region grew dates, and French Sudan, which would eventually become Mali. The festival was heavily modeled on the well-known National Orange Show, which drew attention from media outlets around the country. Boosters hoped that celebrating their unique crop might place them on the tourist map.

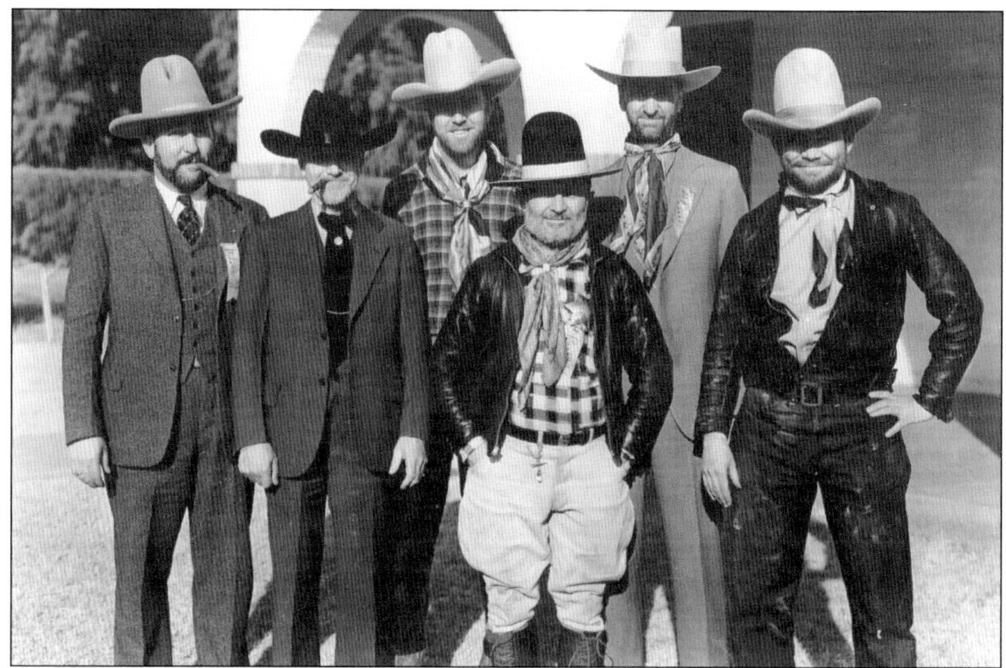

The Indio Civic Club, organized in 1935, oversaw Indio's City (Miles) Park and coordinated nighttime baseball. In 1938, they worked with Riverside County Supervisors to host the first joint Riverside County Fair and Coachella Valley Date Fiesta, taking on heavy financial obligations and donating their time. Here, the Indio Civic Club poses in front of city hall during the 1937 Indio Stampede, a precursor to the revived festival.

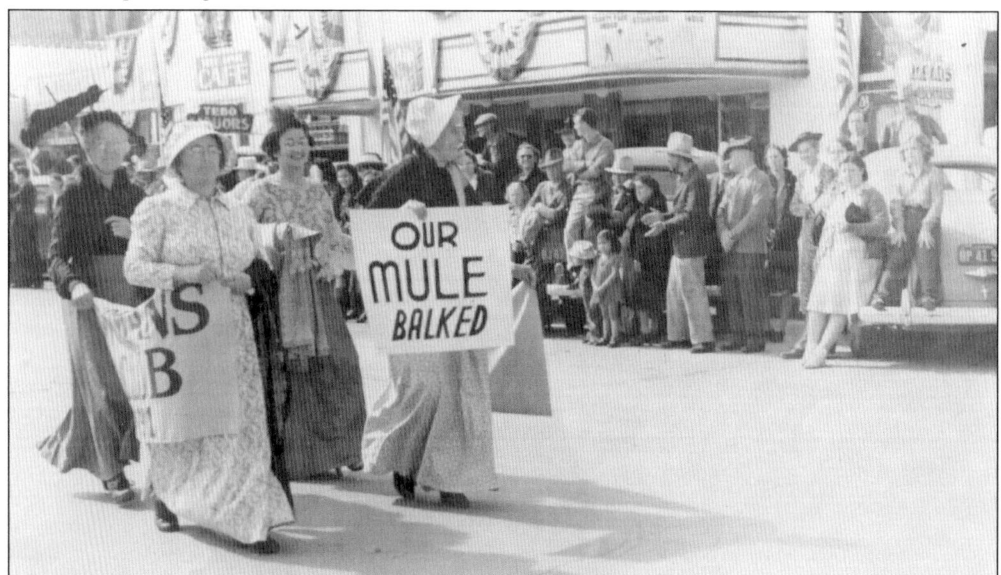

Other local organizations also supported early date celebrations. The Women's Club of the Coachella Valley hosted a Date Fiesta from 1929 to 1931, with a queen, automobile show, carnival, and dance. As the fair evolved, the Women's Club oversaw the queen's costume and her float in the annual parade. Members of the club championed costumes (Western and later Arabian) during fair time, as this parade photograph demonstrates.

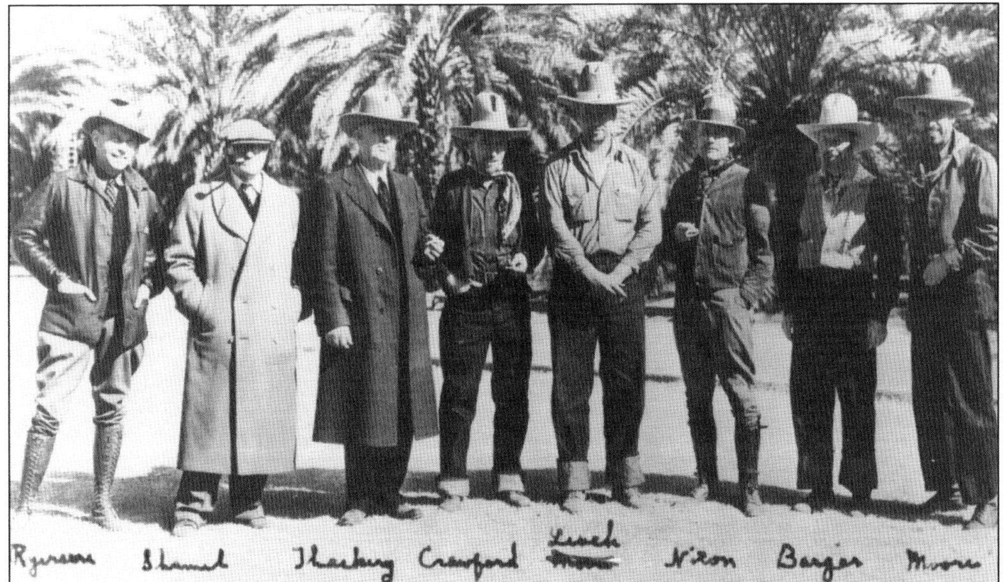

As the fair's Western theme developed in the late 1930s and early 1940s, the desert embraced the cowboy look in conjunction with the rodeos held at the event. Several photographs and newspapers from the time featured male community leaders in their Western garb. This stood in stark contrast with later festivals, where the female body, dressed in the harem-girl fashion, took center stage. While the Middle East was widely feminized in popular culture and associated with the sexuality of the harem, the popular assumptions of the Wild West remained masculine. Above, US Department of Agriculture Experiment Station staff pose in their outfits for the 1937 Rodeo Days. Below, men don their gear in 1938 in front of Ted Johnson's (far right) Shell Station.

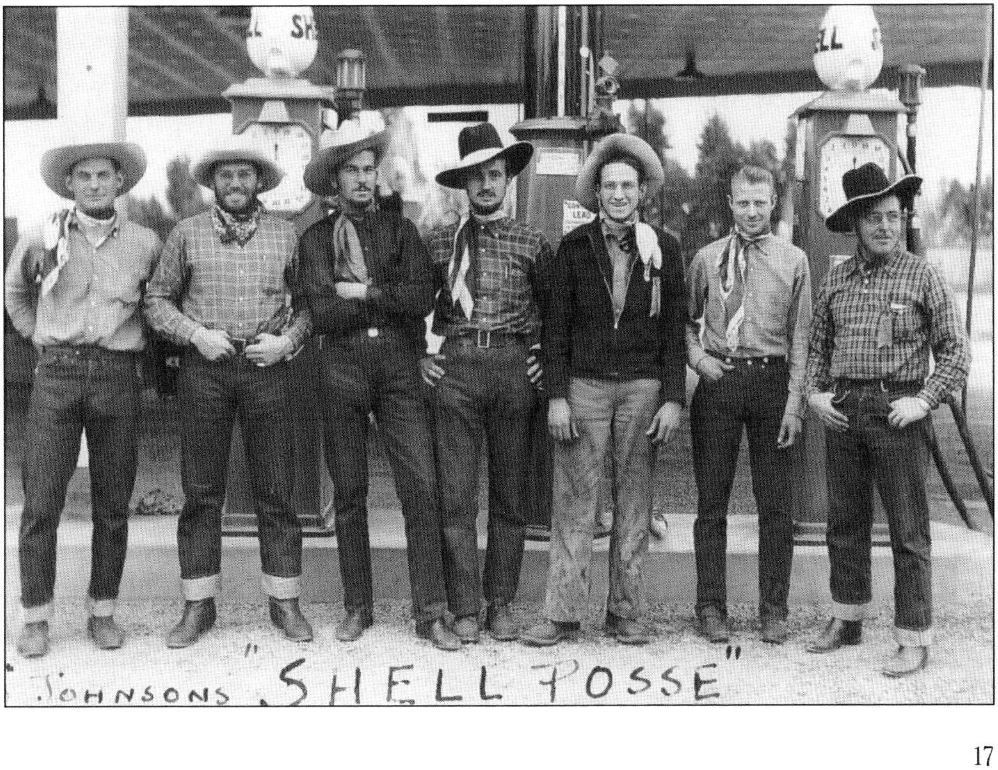

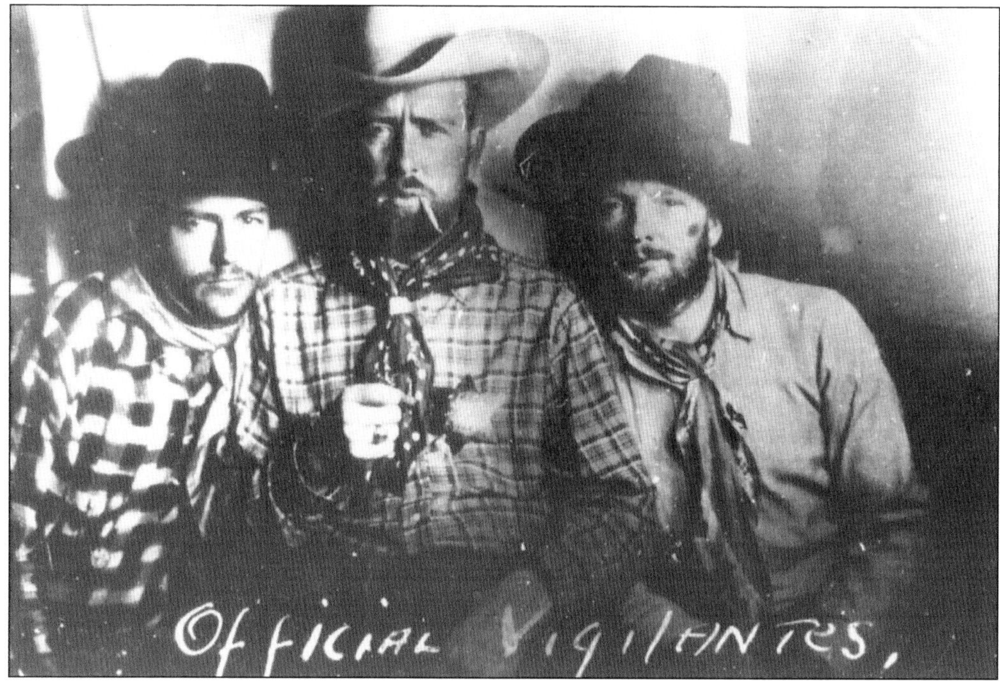

Organizers of the fair hoped to bring attention to the region by encouraging locals to grow out their facial hair in so-called Whiskerino contests. The best and longest beards won special prizes, and Southern California newspapers picked up the story. Prominent community members who wanted to opt out had two options: purchase a "smooth puss" badge that pardoned them from sporting beards or face the wrath of the "vigilantes" who charged them a fine that went to charity. In 1937, when the top photograph was taken, the fine ran upward of $5, and even the chief of police had to pay. In the bottom photograph, nonparticipants were rounded up for comic effect. Kerby Hester stands watch over the offenders, which included newspaperman Ward Grant and city attorney Wallace Rouse.

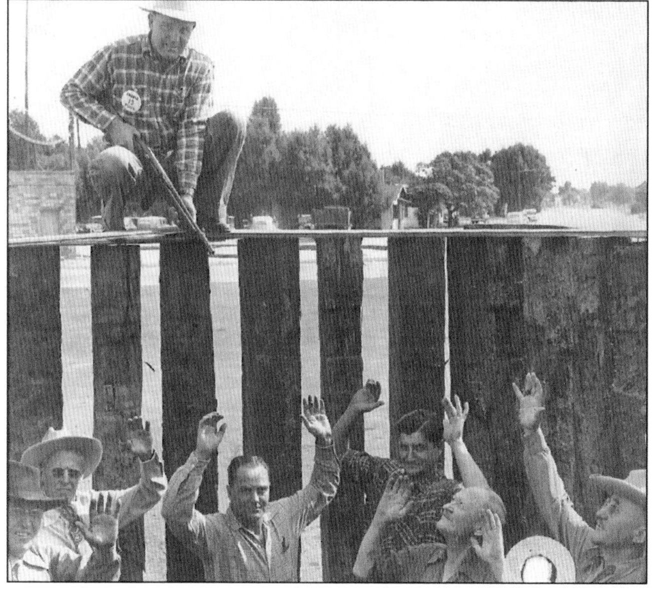

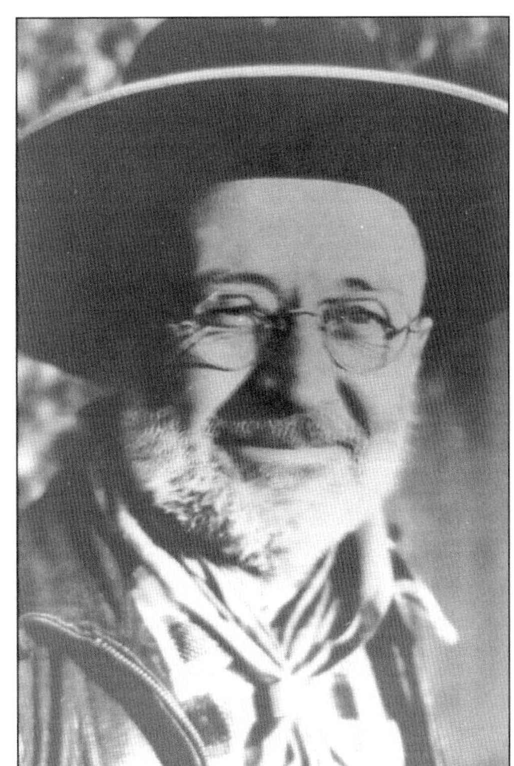

Fairgoers wanted to remember the special event; many took photographs of their Western regalia to commemorate their participation. Frank Tebo, a local businessman, poses at right with his Whiskerino competition beard. One of the rare images of a woman in Western costume shows Betty Riley. Newspapers lamented that fewer women donned cowgirl gear, though local stores advertised Western wear for both women and men.

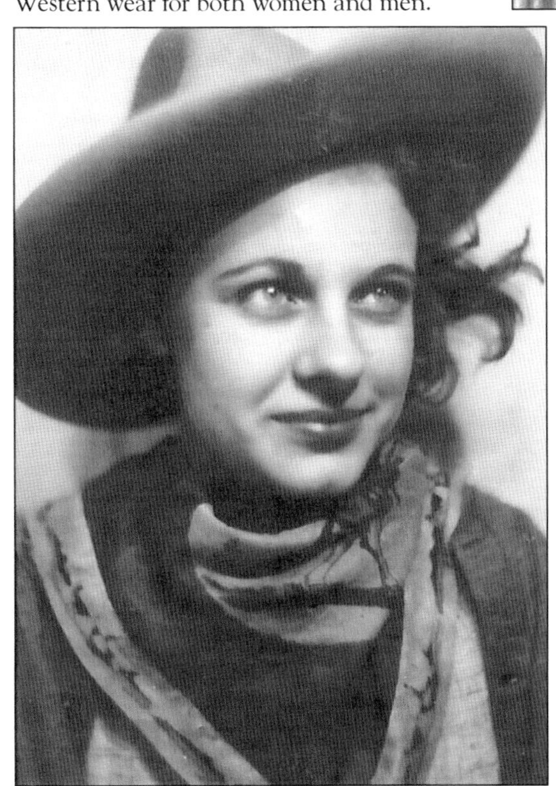

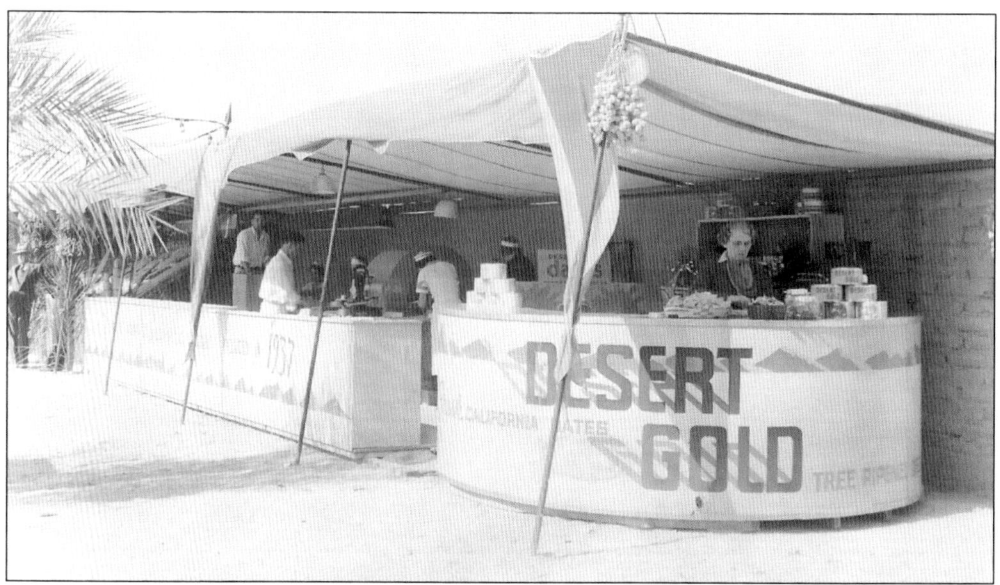

While more recent fairs include a wide variety of non-date-related exhibits and attractions, dates were the star of the show at early festivals. Above, in 1938 a crew manufactured Da-co-nut confections in public view. The treat featured ground dates rolled in coconut and covered with almonds or walnuts. Local growers pooled their dates to participate jointly in the Desert Gold marketing, which aimed at making a united date brand familiar to the American consumer as opposed to small mom-and-pop brands. Sunkist oranges and Sun-Maid raisins developed in the same manner. Below, the first prize-winning California Date Growers exhibit highlights the industry with photographs of groves and a packing plant. Dates made up the gravel for the miniature prospector and donkey, a common sight in the deserts outside of the valley. (Below photograph by Dewey Moore; both photographs courtesy of the Coachella Valley History Museum.)

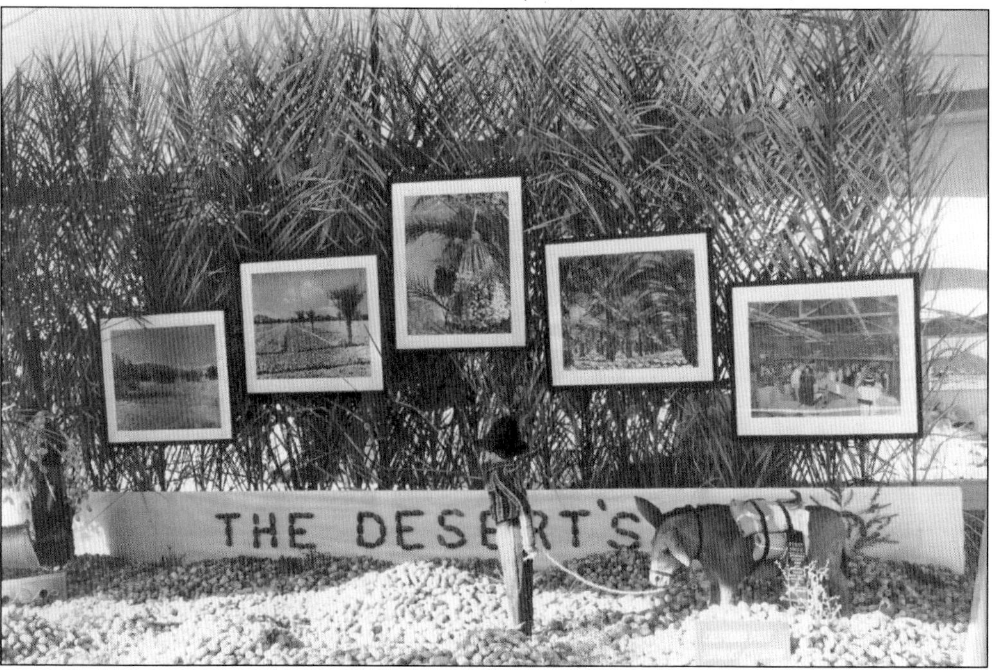

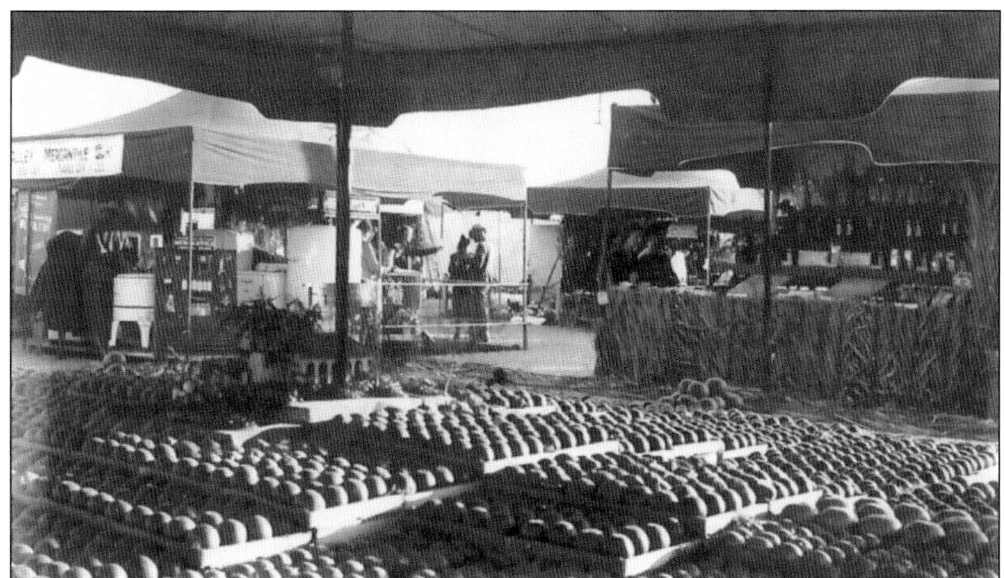

Just as they do today, growers were encouraged to submit their produce for prize money, ribbons, and trophies. Aside from dates, the Coachella Valley was well known for its citrus, especially grapefruit, pictured here. Resourceful organizers used palm fronds to decorate their exhibits. While the date and other crops vied for top billing, locals also displayed goods for sale. The top photograph, for example, shows the Valley Mercantile display, complete with new-model refrigerators and washing machines. The sign listed the store's number so prospective clients would know exactly where to purchase the latest in consumer goods. Since there were no permanent fairgrounds, exhibits shaded themselves with individual covers, below. The Ferris wheel pictured here was a huge draw; the midway shows, which charged a separate fee as they do at today's fair, brought around eight rides to these early festivals. (Photographs by Margaret Tyler.)

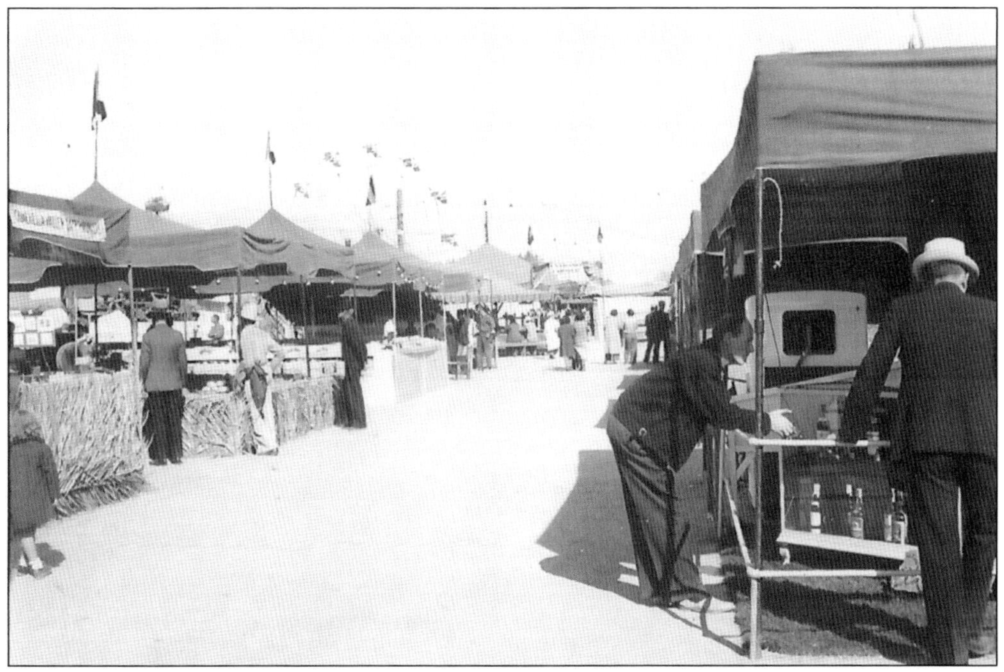

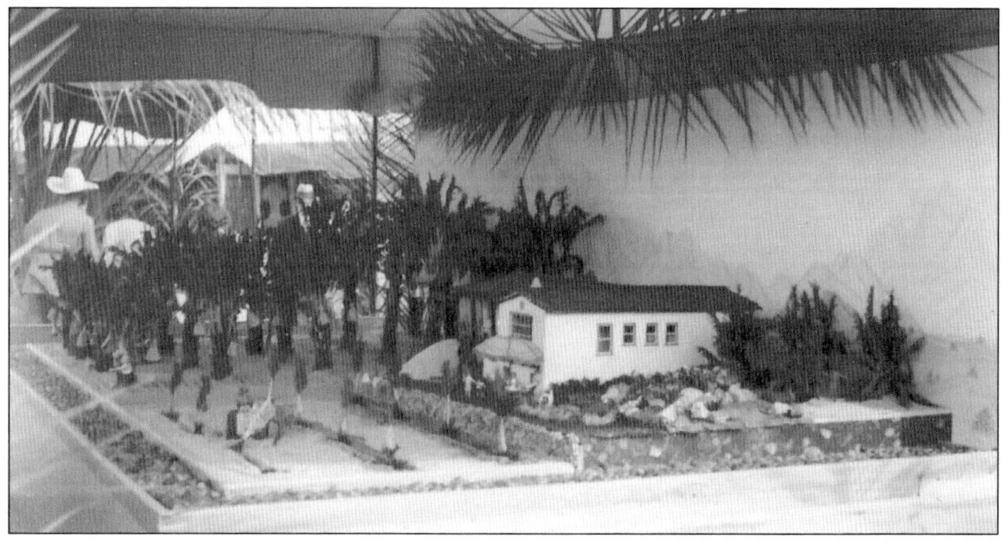

Exhibits from that first festival were incredibly detailed. Above, Dorothy Robinson, sister of date grower Don Mitchell, created this miniature date farm for the Cal-Date exhibit. Crepe paper palm fronds rest atop date bunches made from strings of wheat. Even the backdrop features the local mountains. Below, note the *palmero*, or date-palm worker, in the bottom right corner. In the early years, most trees were lower to the ground. This meant date labor could be completed easily by family members of growers. As the trees rose, working conditions became more dangerous. During World War II, much of the local labor force went to war or relocated to more urban areas for better-paying defense jobs. The bracero program filled the need for agricultural workers, encouraging the tradition of Mexican and later Mexican American palmeros. (Photographs by Margaret Tyler.)

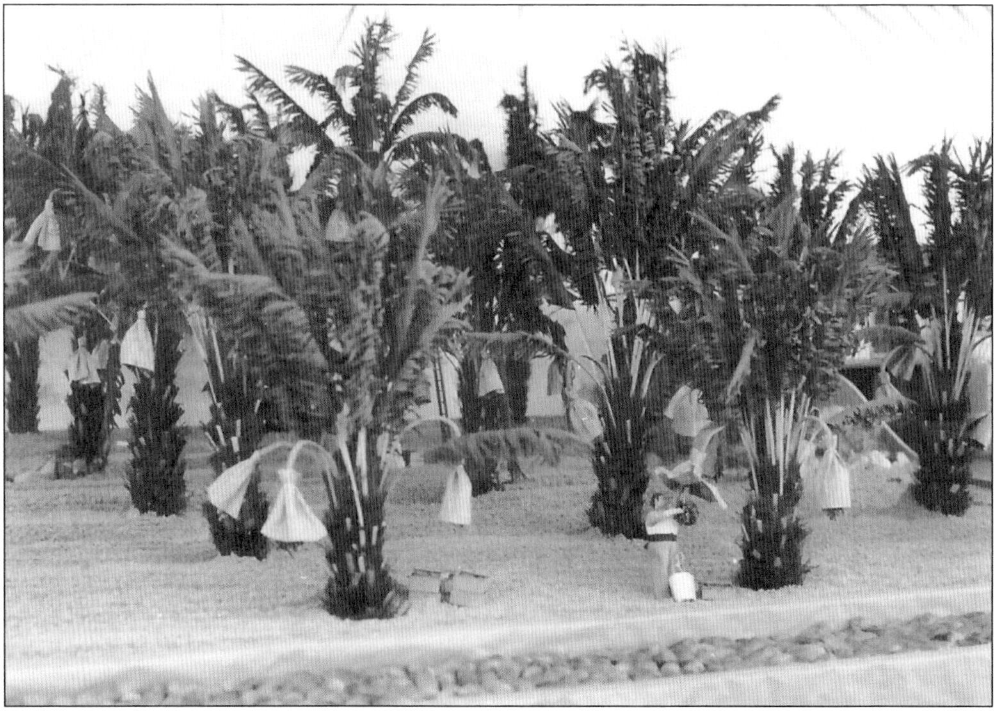

The Coachella Valley Date Growers display their wares in this exhibit from 1938. A multitude of products, including date-plum pudding, date bars, date paste, Shield's Date Garden's famed date crystals, and treats from the Garden of the Setting Sun, drew consumers' attention. Visitors to the area often picked a favorite date shop and wrote annually to purchase these products as Christmas gifts. (Photograph by Margaret Tyler.)

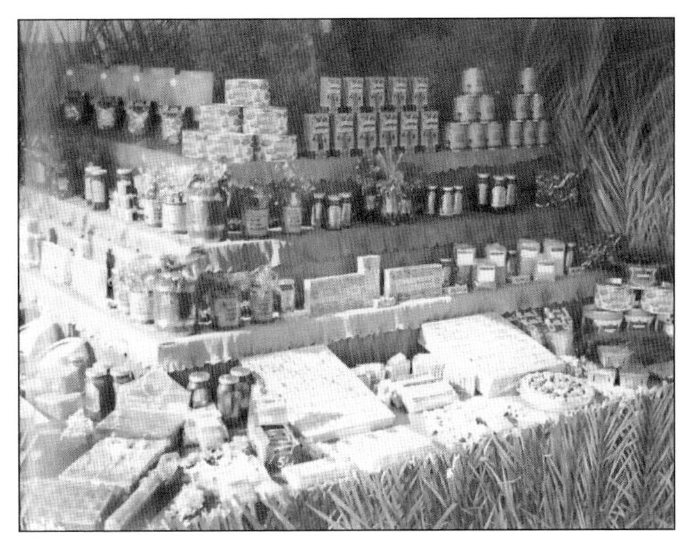

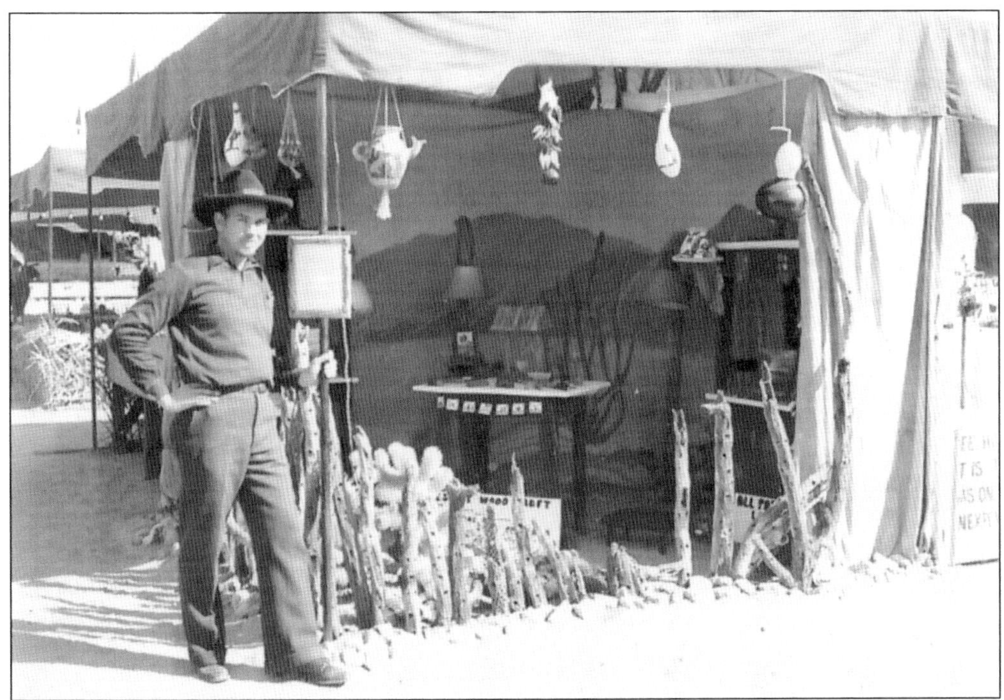

Frank Freeland poses with a display of woodcraft made from ocotillo. Native to the Coachella Valley, the plant's carcass leaves the hole-ridden staffs seen here. Ocotillos, which have long rows of sharp spines, provided the inspiration for barbed wire. The painted backdrop includes an image of a living ocotillo. Its bright-red blooms are popular with tourists. (Photograph by Margaret Tyler.)

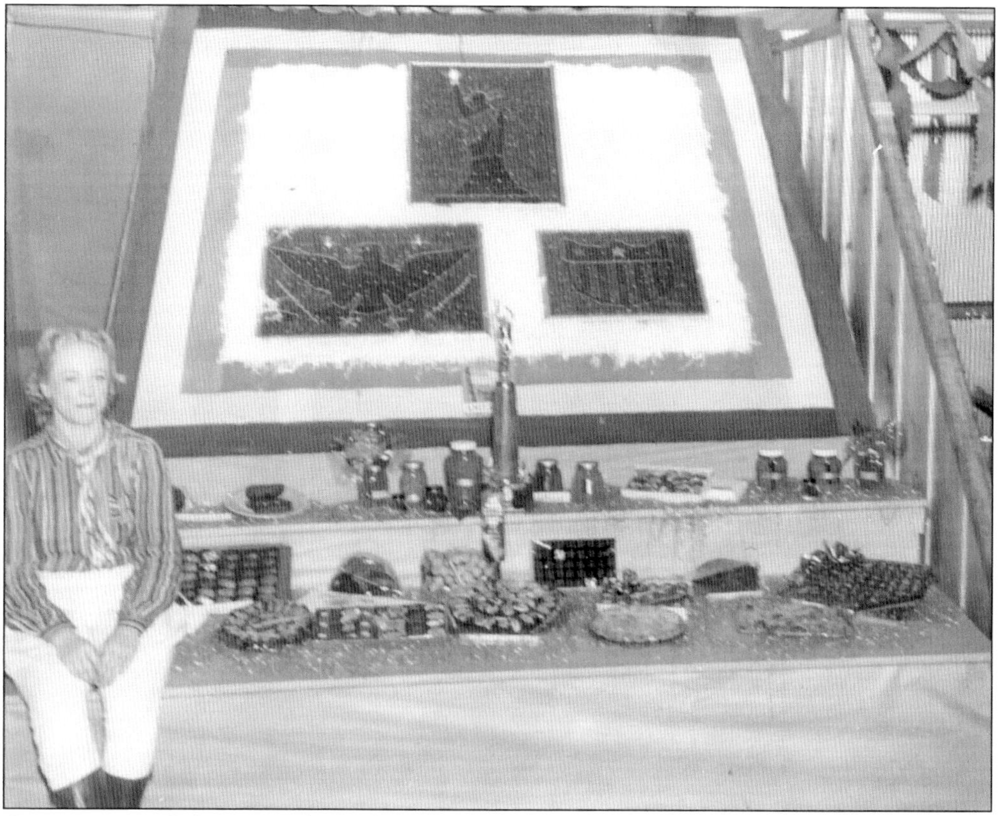

Creating art with fruit proved popular at fairs around the state; this date artwork demonstrates the practice. The prize-winning exhibit from 1941 features iconic American symbols such as the eagle and Statue of Liberty, reflecting a prewar patriotism. A trophy can be seen in the center, amid homemade date confections. A cowboy date "painting" sits centered in the bottom photograph. While the Arabian theme marketed dates in local shops and for nationwide importers, occasionally, growers turned to a Spanish theme to sell their wares. The Desert Date Shop exhibit won first prize in the 1939 festival. Hand-painted gift boxes feature Mexican stereotypes common in Southern California imagery, including a man with a sombrero sleeping under a date palm tree and a dancing senorita.

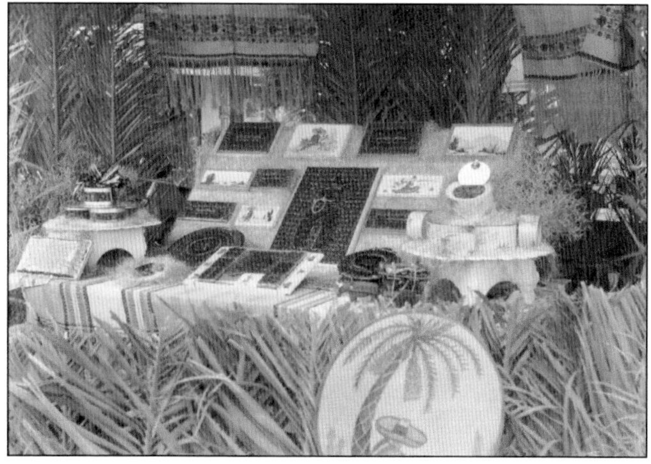

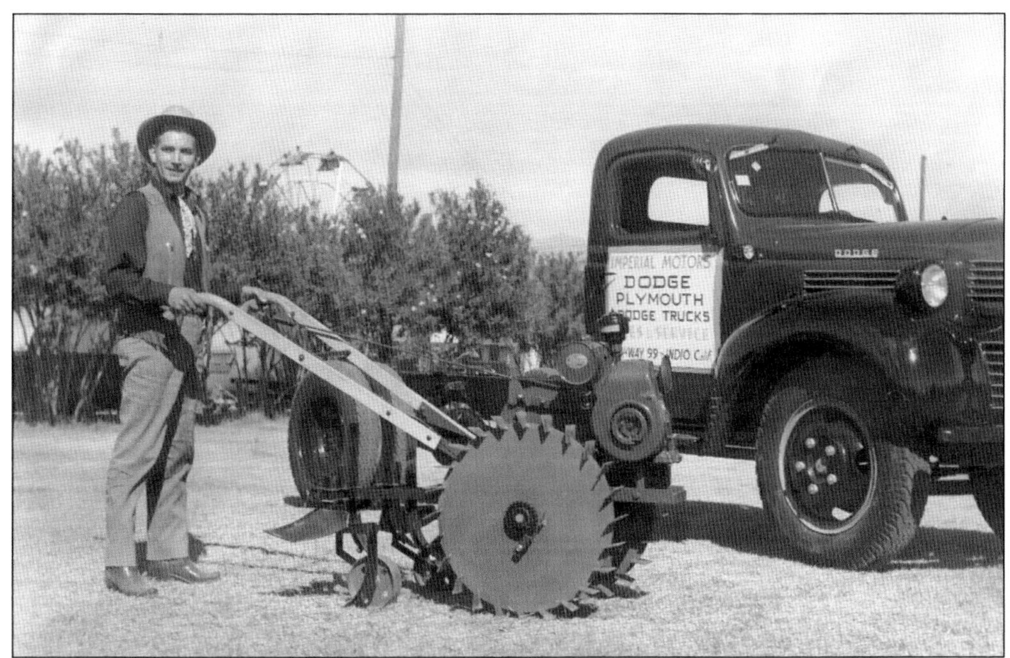

Community businesses appeared regularly at the fair. While this truck from Imperial Motors no doubt was driven in the parade, local newspapers reported cars on display as well, fitting in with the consumer nature of the event. Farm equipment was also exhibited at this agricultural festival, which drew farmers from around the county. Note the Ferris wheel in the background.

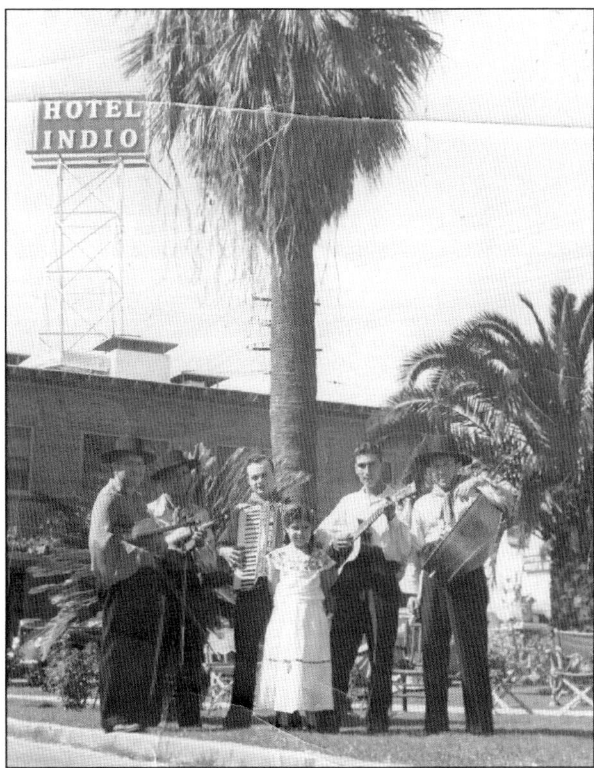

The 1941 festival listed Arturo Preciado and the Golden Gate Expo Orchestra as entertainment. Preciado lived in Pomona but played at fairs in Los Angeles and Palm Springs. His children recall that Preciado and his band, the Troubadours, also paraded around the fairgrounds after 1947. Here, the Troubadours pose in front of the Hotel Indio with Preciado's daughter Priscilla Garcia, who would later become a festival princess. (Courtesy of the Preciado family.)

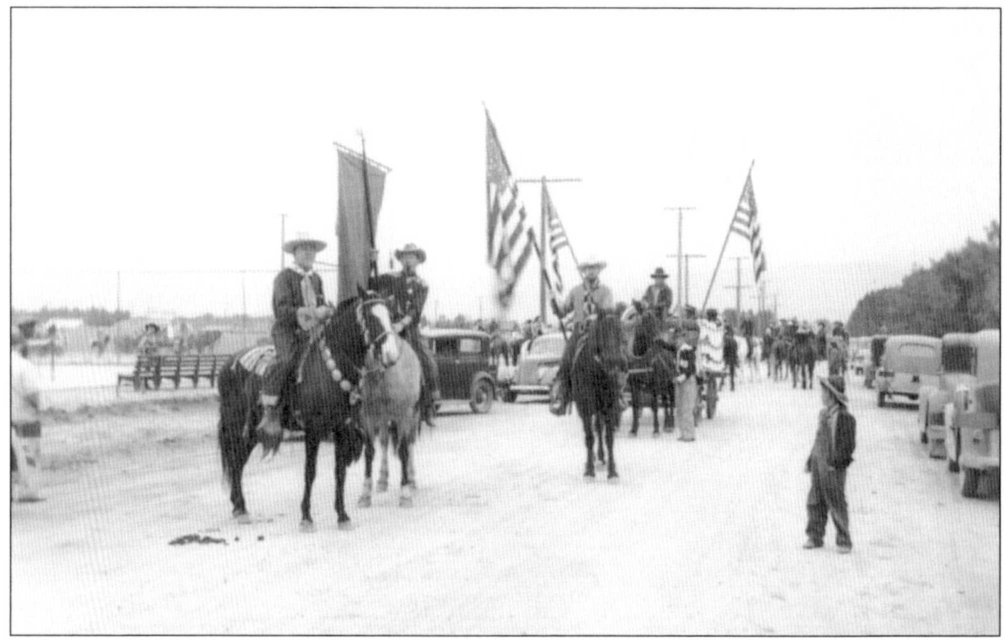

The 1938 grand marshal was the sheriff of Riverside County, Carl Rayburn. Traditionally, local law enforcement led parades alongside the grand marshal; in this case, they were one and the same. This image not only shows the Indio sheriff's posse but also highlights Indio's City Park, with its fenced tennis court to the left. The participants are lined up along Deglet Noor Street, unpaved at the time.

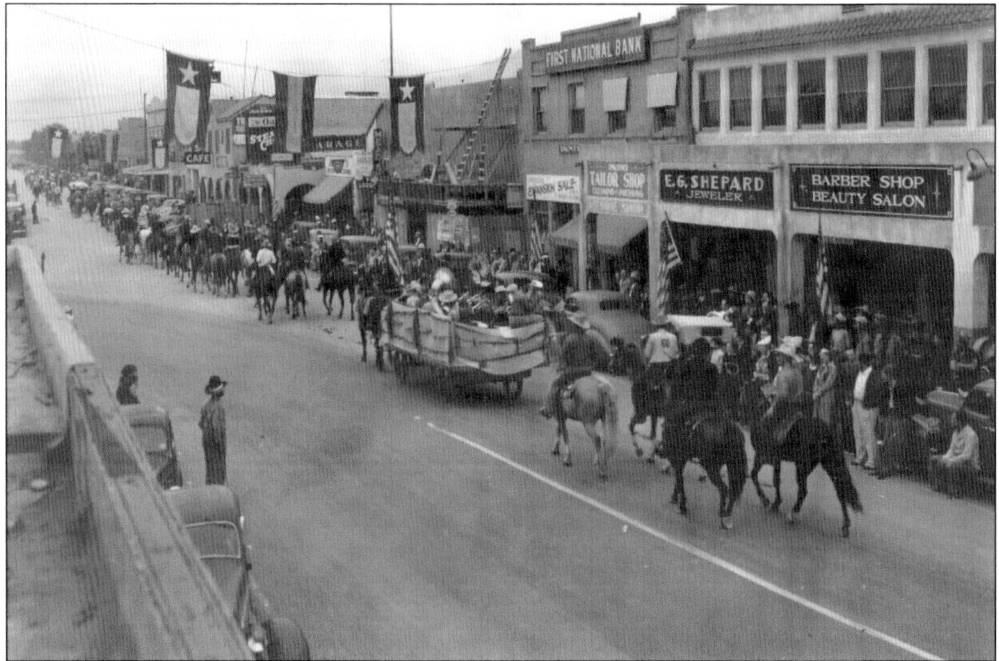

The parade was a crowd favorite at these early festivals. More than 500 mounted entries appeared in the 1938 parade. Both spectators and parade participants donned Western garb in this 1939 photograph, while a band performed in the horse-drawn wagon.

Two

Fairgrounds

Caravansaries and Old Baghdad Invoked

When the Riverside County Fair came to the Coachella Valley, it needed a more permanent home than Miles Park. In 1941, the county purchased the current fairgrounds for $10,000. The county received matching funds for their fair investments through funding provided by pari-mutuel horse wagering. The location, just off of today's Highway 111, was at the time close enough to Highway 99 to be convenient but far enough away not to impede traffic. The Gonzalez family sold the 40-acre ranch property. Prior to the sale, they grew citrus and Deglet Noor dates, imported from Algeria, on the property. Younger family members parked fairgoers' cars on their adjacent land to earn extra money to spend at the fair. The going rate rested at $1 for cars and $2 for trucks.

Building and improvements on the fairgrounds mostly stalled until after World War II, though a portion of the land was leased to the government for railroad-worker housing. A main exhibit hall, restroom, and a small segment of the fence were the only structures built in time for the 1947 fair. By then, an Arabian theme had been chosen. Former Hollywood set designer Harry Oliver devised a Moorish ornamental wall, ticket offices, and the Old Baghdad Stage, all of which still exist today. While the Arabic design of the stage drew attention, its status as an outdoor amphitheater did as well. Where else in the United States could one hold an outdoor pageant at night in February in such a mild climate? Eventually, the fairgrounds housed a large arena for horseshows and camel racing. The Promenade of Palm Trees featured different varieties of dates while large Arabian-themed billboards, signs, and even statues beckoned visitors to the most unique fairgrounds in the nation.

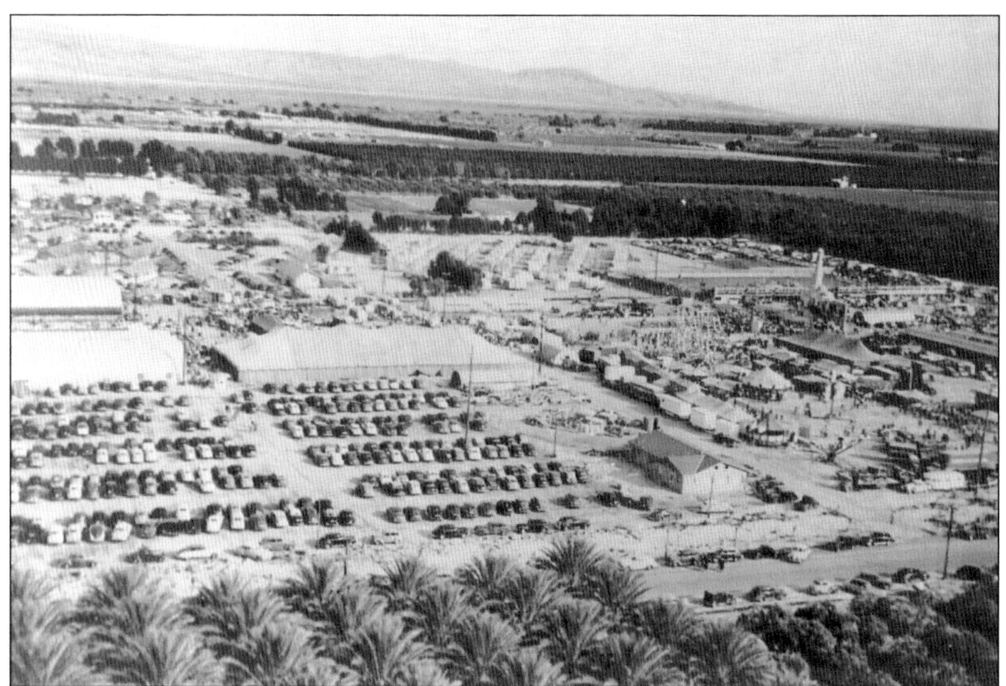

Though surrounded by housing and businesses now, in the early years, the festival sat in the middle of agricultural land, mostly dates and citrus. This photograph from the early 1950s includes the carnival rides and parking lots. Behind the fair rests the railroad-worker housing.

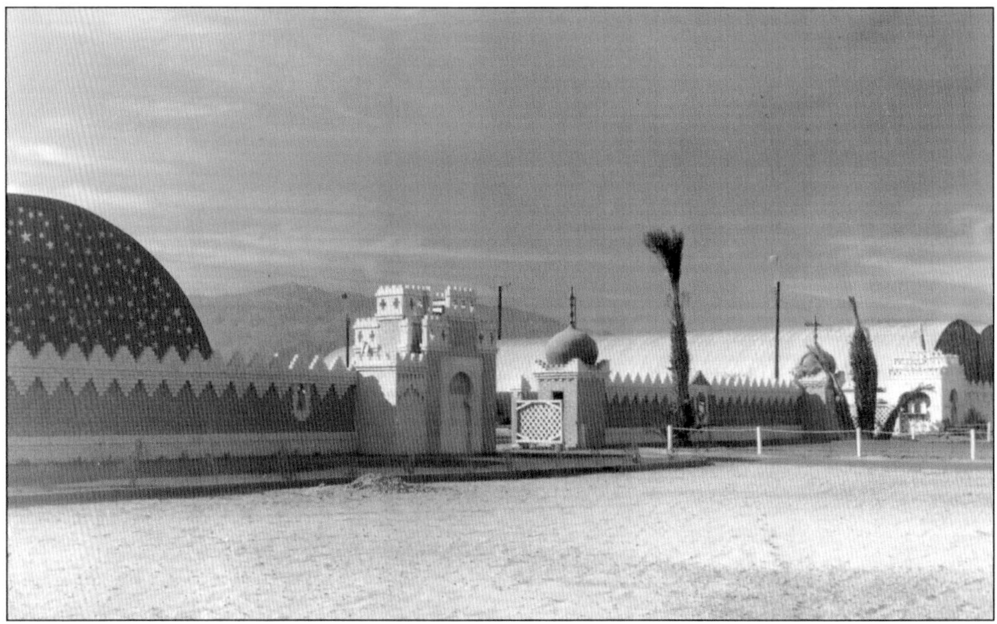

When organizers began building the fairgrounds after World War II, they faced a shortage of supplies. Manager Robert Fullenwider recalled seeing an image of curved Quonset huts in Damascus. Widely used by the US military in World War II, these huts were cheap and quickly assembled. They not only solved the supply issue but also, in the eyes of Fullenwider, fit into the Arabian theme, as seen in this 1955 image.

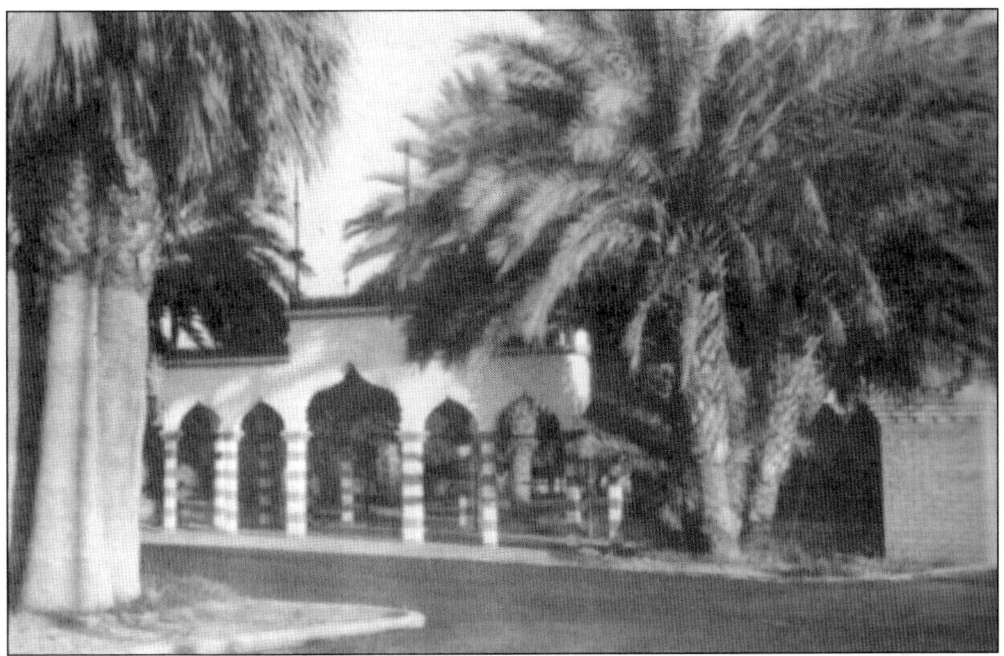

The main entrance signaled to visitors their journey to a foreign land was about to begin. Designed by Harry Oliver, the gate and ticket booths welcomed fairgoers as they stepped into America's Arabia. Both native *Washingtonia filifera* fan palms and date palms invoke lush oases. (Author's collection.)

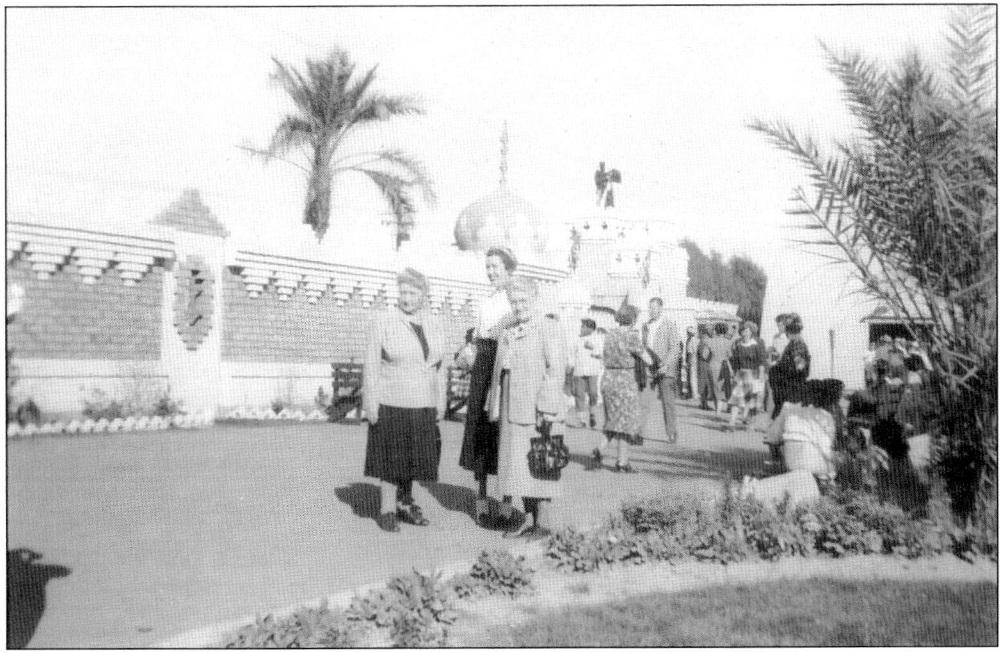

Architecture brightened the Arabian theme but so, too, did human engagement with it. Costumed guards patrolled the wall at the front entrance, conjuring a scene from a Middle Eastern palace. One can be seen in the middle of this 1949 image. (Courtesy of Linda Beal and the History Hunters.)

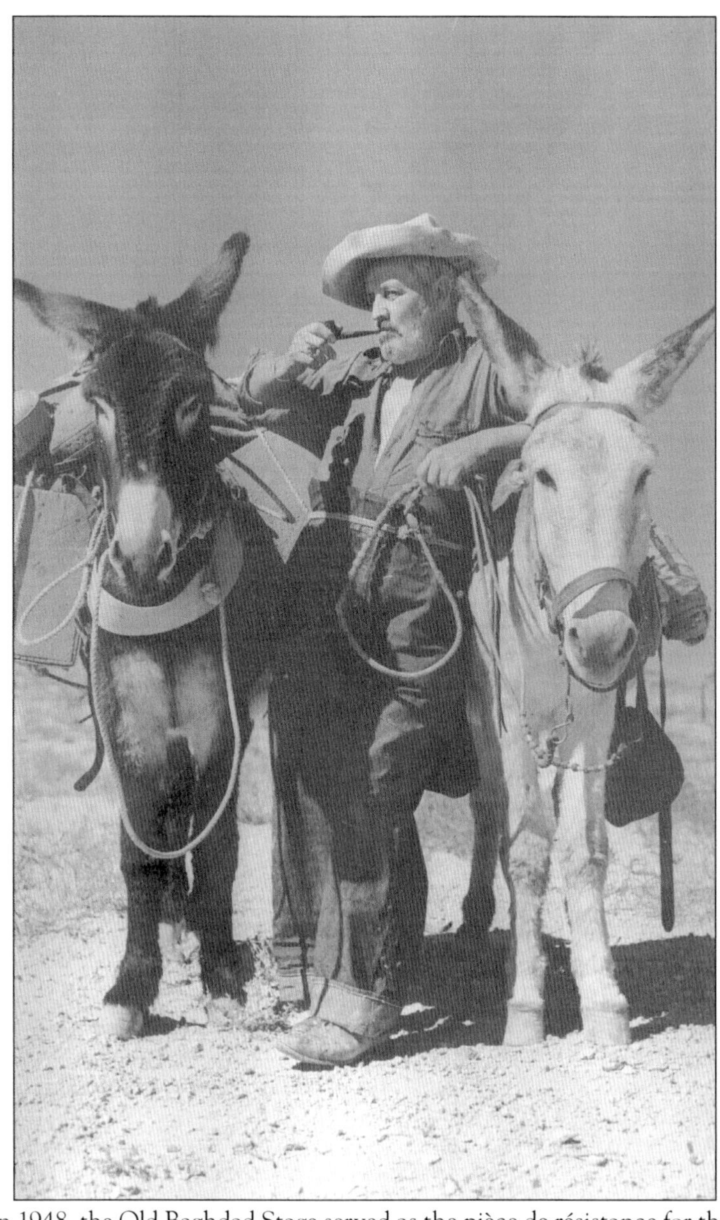

Completed in 1948, the Old Baghdad Stage served as the pièce de résistance for the fairgrounds. Harry Oliver designed the stage with a scale model rather than blueprints, showcasing the model at the California State Fair to high acclaim. Known regionally as a humorist, he already had experience with whimsical design, creating Beverly Hills Witch (Spadena) House, the first iconic Van De Kamp Windmill building, and the Gold Gulch concession at the California Pacific International Exposition held in San Diego in the mid-1930s. Nominated for an Academy Award in 1928, Oliver worked as a set designer and artistic director for *Ben Hur, The Good Earth, Viva Villa,* and *Street Angel.* After adopting the persona of a desert rat, Oliver built Old Fort Oliver in Thousand Palms, writing desert-color stories for newspapers and even *Life* and *Desert* magazine. He also produced the *Desert Rat Scrapbook*, a beloved periodical in the Coachella Valley. Oliver championed secession from Riverside and San Bernardino Counties to create Desert County and was well known for his tall tales. (Courtesy of the Palm Desert Historical Society.)

Designed to look like a caliph's palace in medieval Baghdad, the open-air stage has hosted the nightly musical pageant based on *One Thousand and One Arabian Nights* since 1948. While the stage's front drew acclaim with the show, the market-themed rear stage was a favorite for publicity photographs. The details included arched doorways and *mashrabiyas*, wooden latticework-covered windows, a tradition in Middle Eastern architecture. The Arabic on the left reads "Pastry Shop" while the writing on the right reads "Dates Shop—The Only One in America." The middle inscription is a Farsi translation of the English below it ("Arabian Foods & Sweet Meats"). In the fair's early years, the public was encouraged to take photographs, often alongside a camel or a queen contestant in costume. (Right, courtesy of Jeanne, Jerry, and Jennifer Blair; below, courtesy of Linda Beal and the History Hunters.)

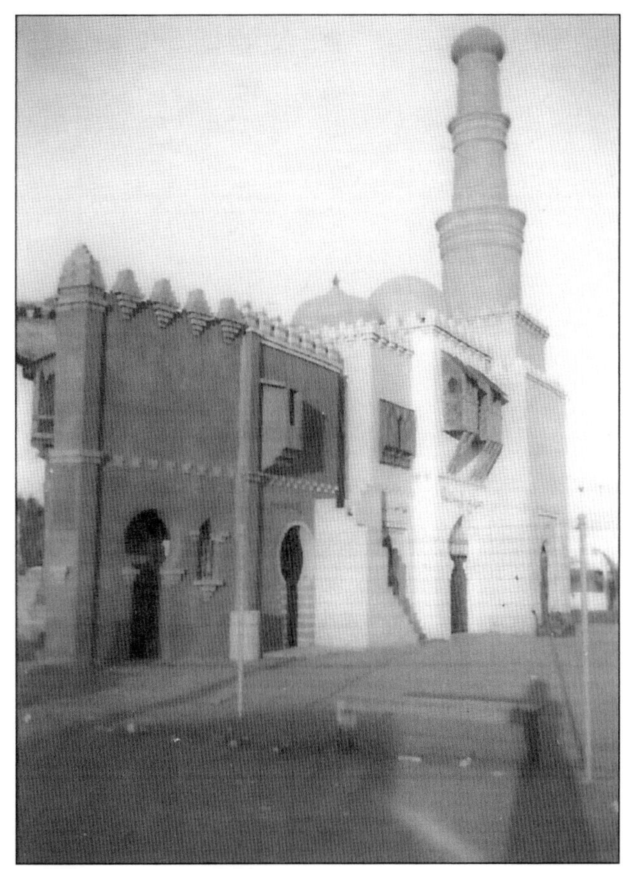

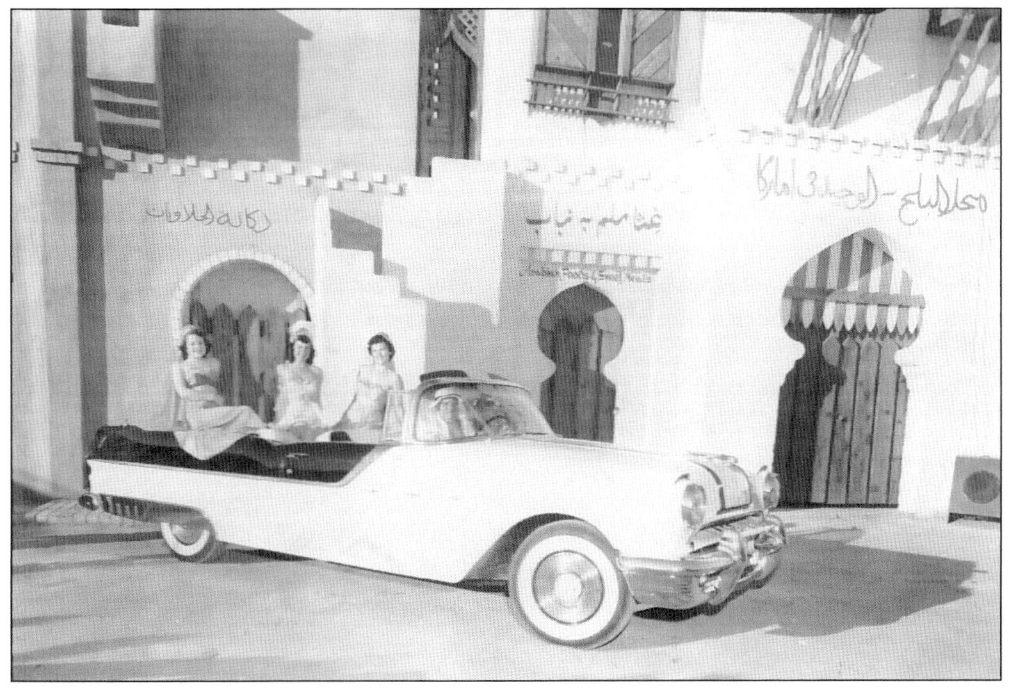

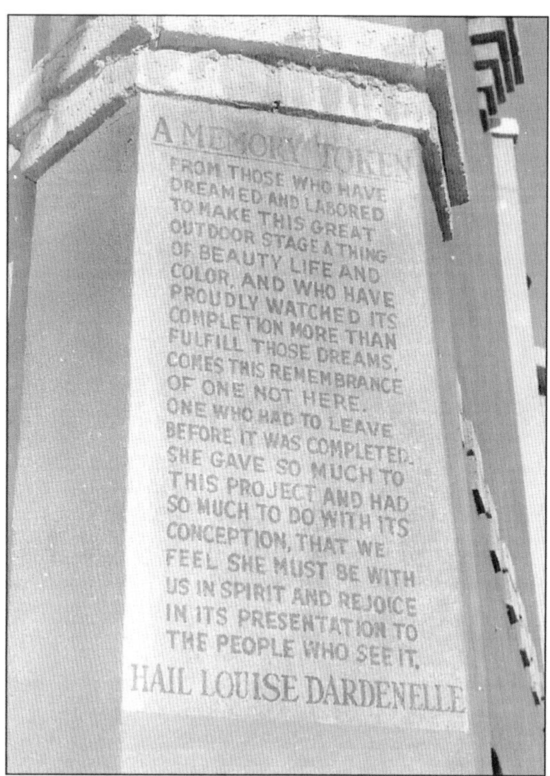

Louise Dardenelle conceived the *Arabian Nights* pageant in 1947, though she passed away in December of the same year, unable to see her work come to fruition. As a tribute to her, fair leadership inscribed this message into a wall of the Old Baghdad Stage. Prior to her death, Dardenelle published a science-fiction novel, earned a doctoral degree in music, and opened the Conservatory of Music, Art, and Dancing in Indio.

Attendees today watch the pageant from folding chairs, but for some time wooden benches lined the area in front of the Old Baghdad Stage. This image, taken from the stage's minaret, also shows the unpaved nature of the fairgrounds and a "Venus Dip" attraction (at center left), a fundraiser for the Coachella Valley Lions Club.

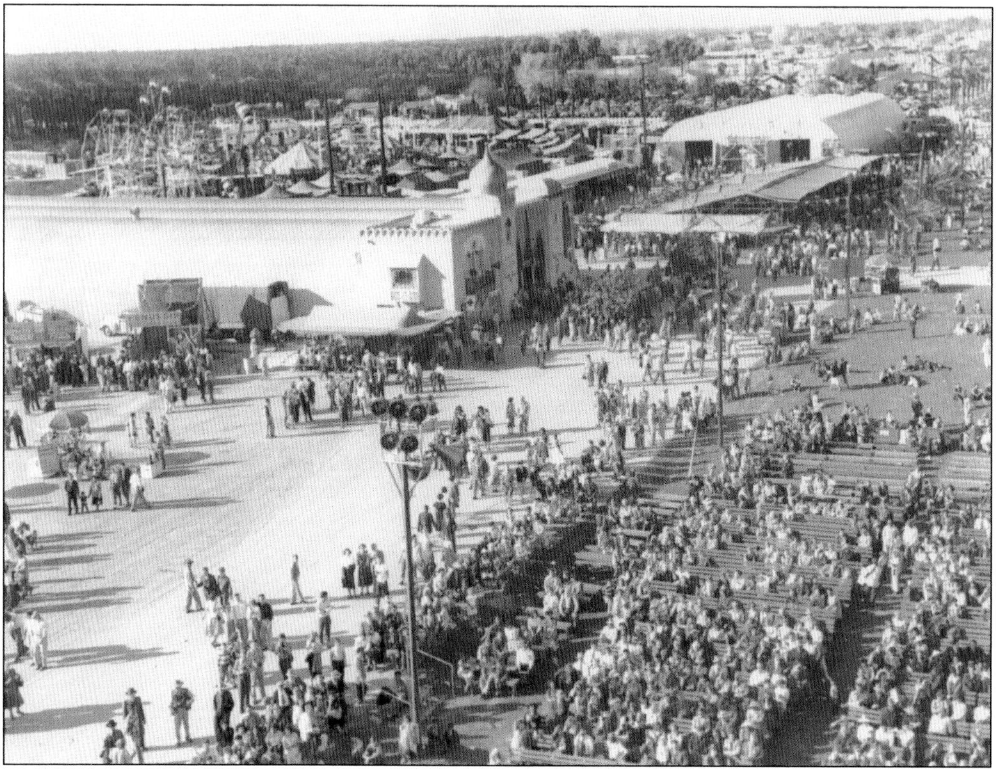

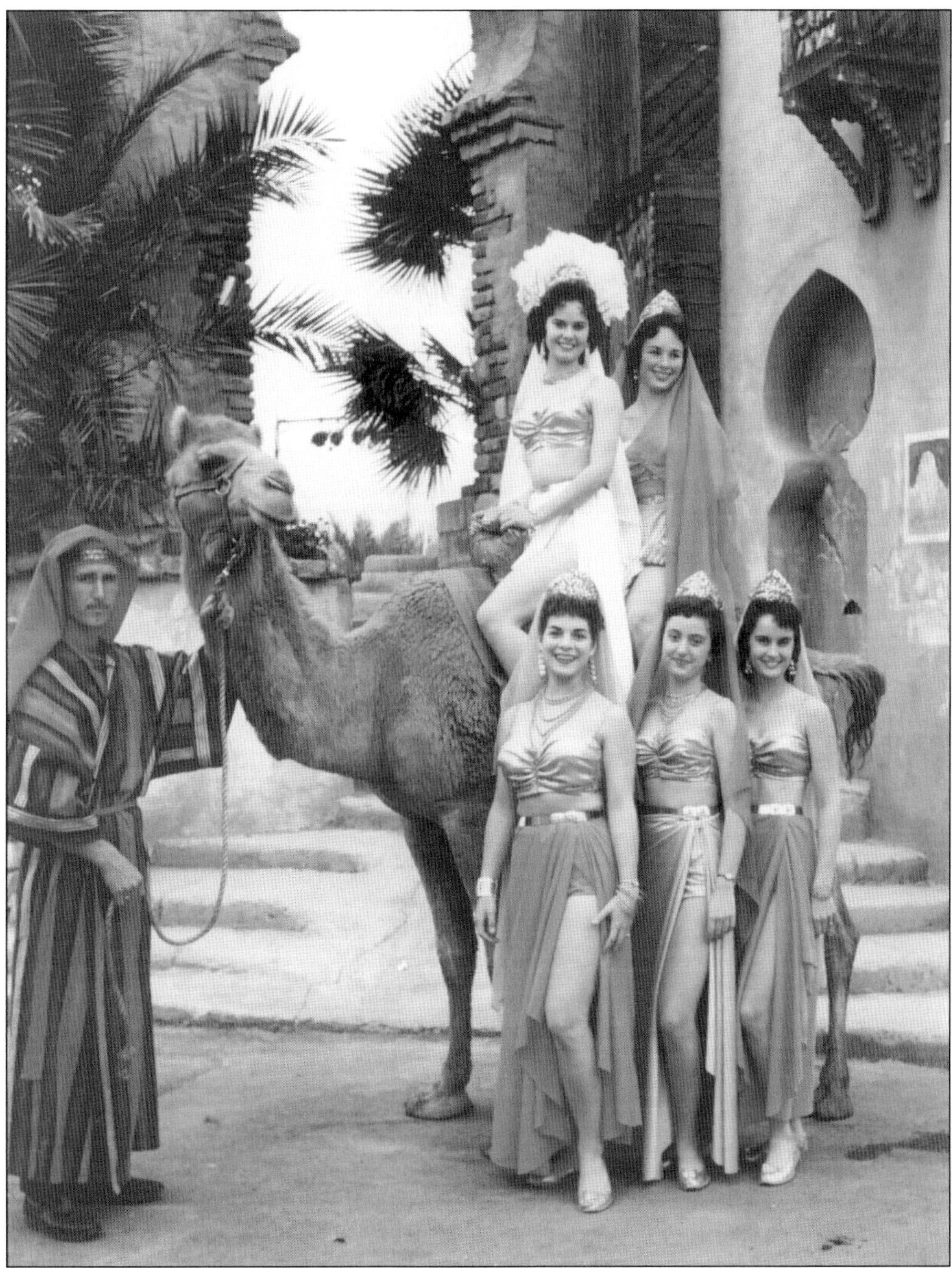

For viewers with a little imagination, the fairgrounds appeared to be Middle Eastern ruins. At least, that is what organizers hoped. The Old Baghdad Stage, in particular, saw countless photo shoots, with images reproduced in everything from *Life* to the *New York Times* and even *Hot Rod* magazine. The side of the stage pictured here incorporated landscaping as well, implying a lush palace garden. Harry Oliver, a former Hollywood set designer, relied on details to set the stage apart; painted cracks and artificially weathered steps invoked an ancient palace.

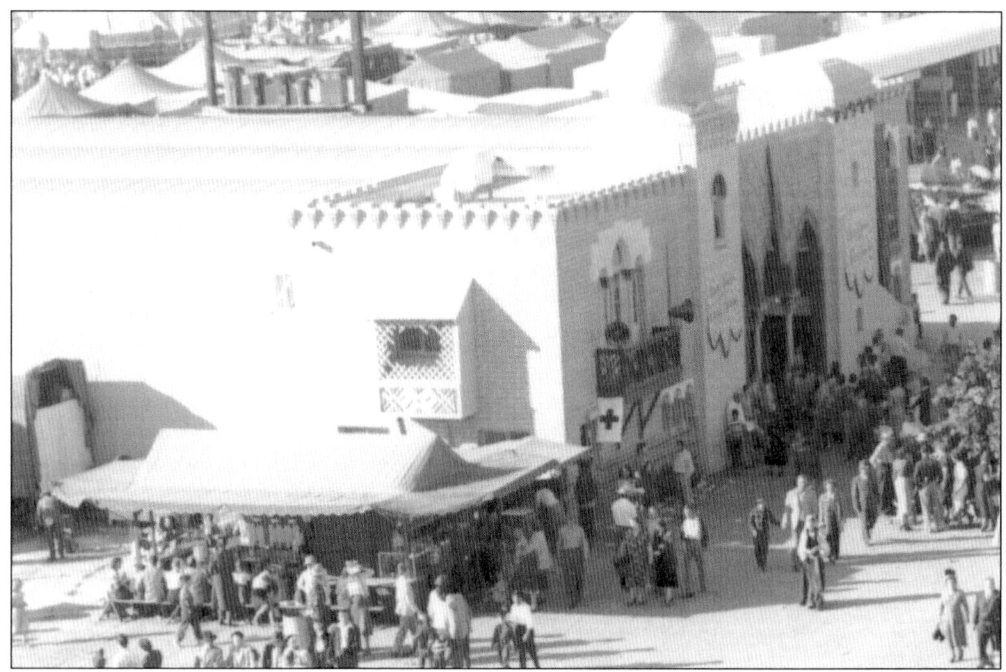

Completed in 1950, the Exhibit Bazaar Building, later known as the Palace of Commerce, combined two building trends at the fair: Moorish-inspired architecture and the Quonset hut. The structure housed fair administration, the first-aid station, and a girls' dormitory for the queen contestants. Replaced by the Fullenwider Building in 1990, both buildings hosted commercial exhibits that sold everything from appliances to jewelry.

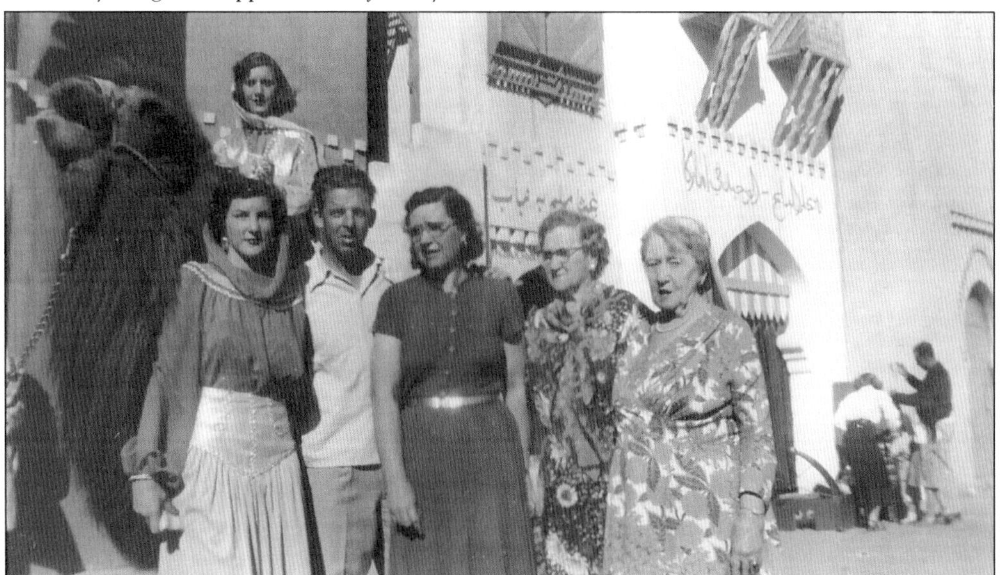

Arabic peppered the fairgrounds, as organizers believed it added authenticity. Historically and today, few residents of the Coachella Valley speak Arabic or have family ties to the Middle East. The Peters family, which ran several successful shops and a restaurant, proved the exception. Marshall Peters, born in Syria, wrote the Arabic words on the fairgrounds, seen here behind the Old Baghdad Stage. (Courtesy of Linda Beal and the History Hunters.)

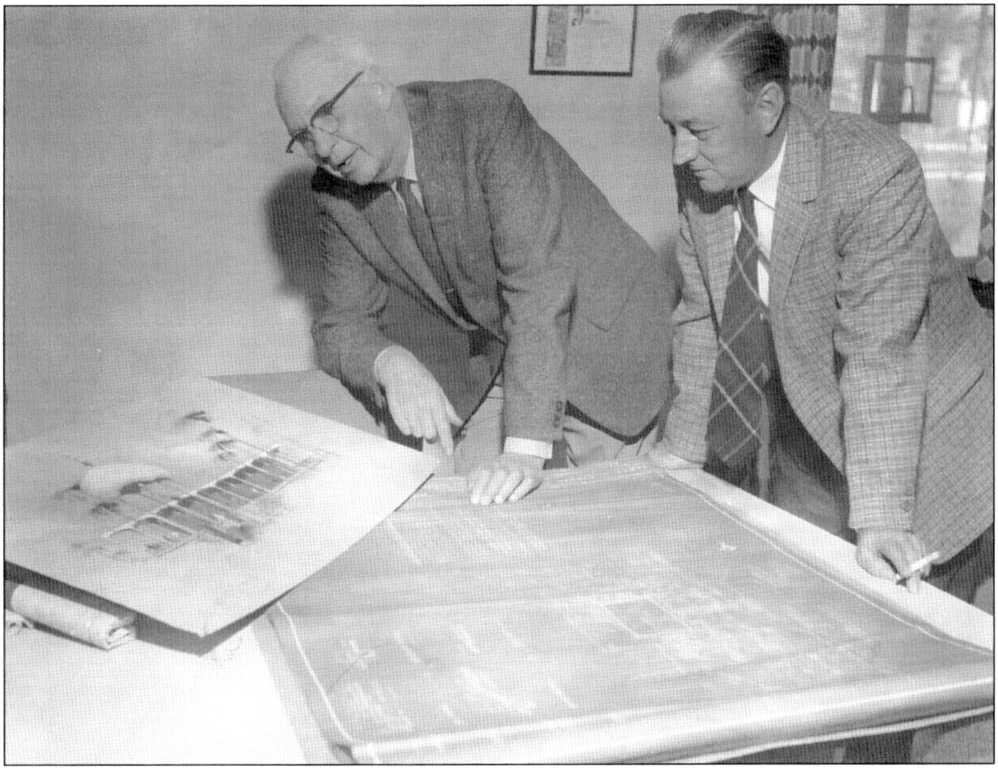

Completed for the 1962 fair, the Caravansary housed a 260-seat cafeteria. Above, fair manager Robert Fullenwider (left) and assistant manager (later manager) Charles Wameling review blueprints for the building prior to construction. Eating holds a special place in most people's memories of the fair. In early years, clubwomen often cooked for large crowds, using the event as a fundraiser. In 1978, the Caravansary served food year-round, hosting banquets and wedding receptions as well. Today, most people purchase food from the concessionaires, and the Caravansary houses the fine art exhibits. The mushroom-like awning invokes the Middle East with its crescent moon but plays with modern architecture as well. (Above, courtesy of Linda Beal and the History Hunters; below, author's collection.)

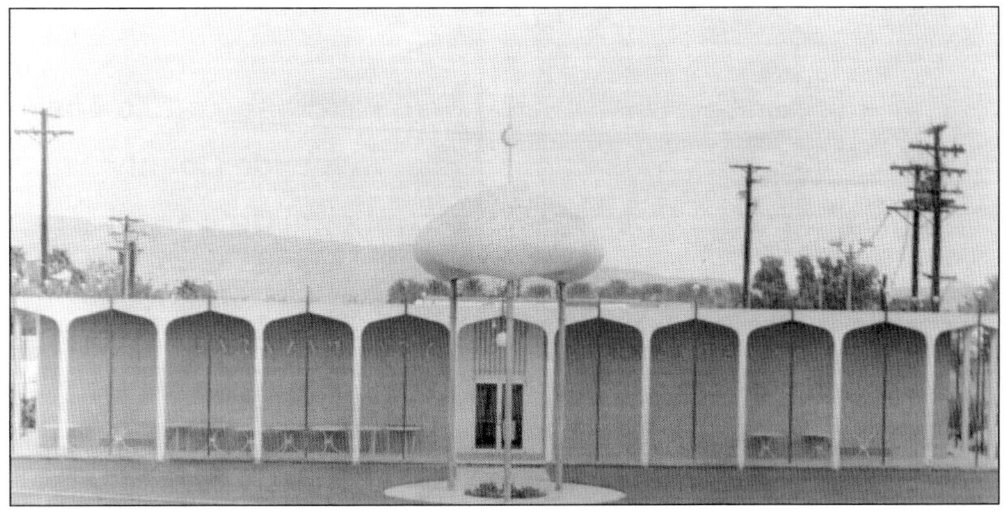

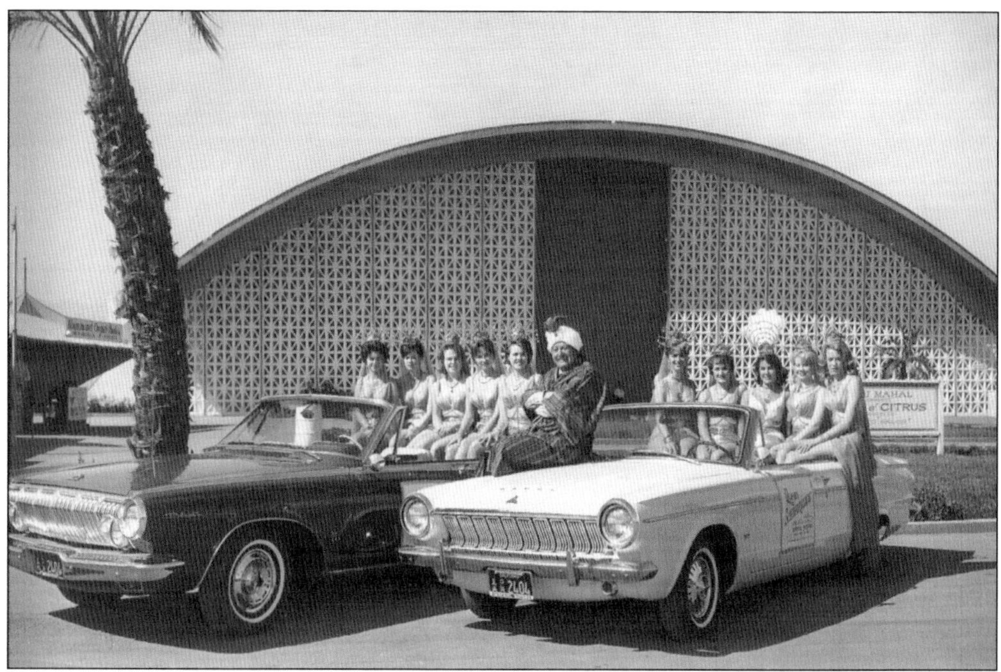

The Taj Mahal Building has long held agricultural exhibits during fair time. The original Taj Mahal in India sprang to life during the Mughal reign, from royalty with heavy ties to Persia, thus its design influences from the Middle East. However, its presence at the fairgrounds also reflects a larger practice of Orientalism, blending vastly different spaces and cultures together under the blanket descriptor Arabian, or Oriental. Above, the white brise-soleil suggests the modern architecture popular during its unveiling yet references the mashrabiya latticework common in the Middle East since medieval times. Below, in 1996, the fair commissioned Bill Weber to paint a mural of the Taj Mahal, bringing India to Indio and beginning the city's famed mural projects. (Above, photograph by Roy Gillman; below, courtesy of Amanda McCormick.)

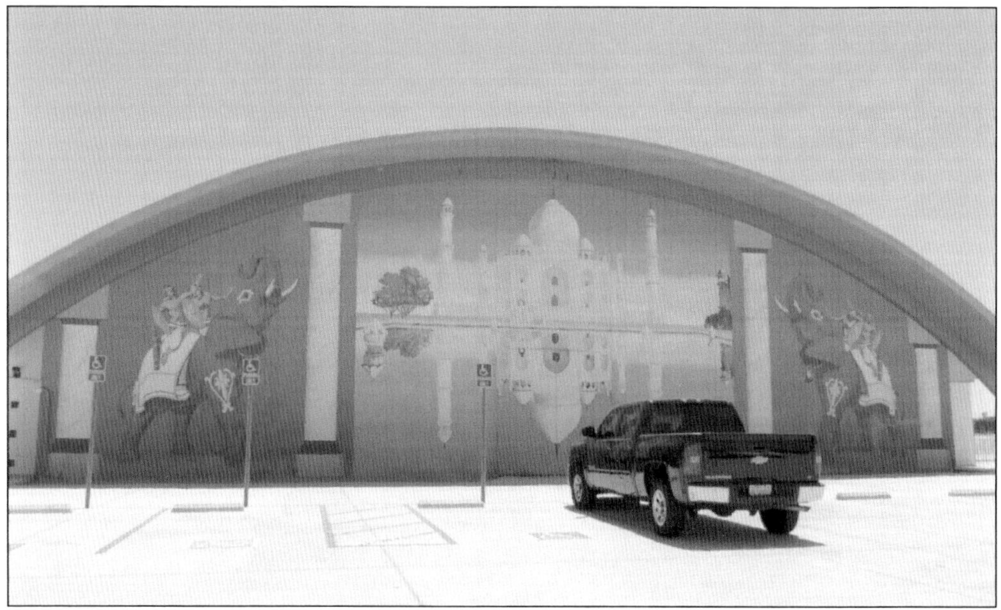

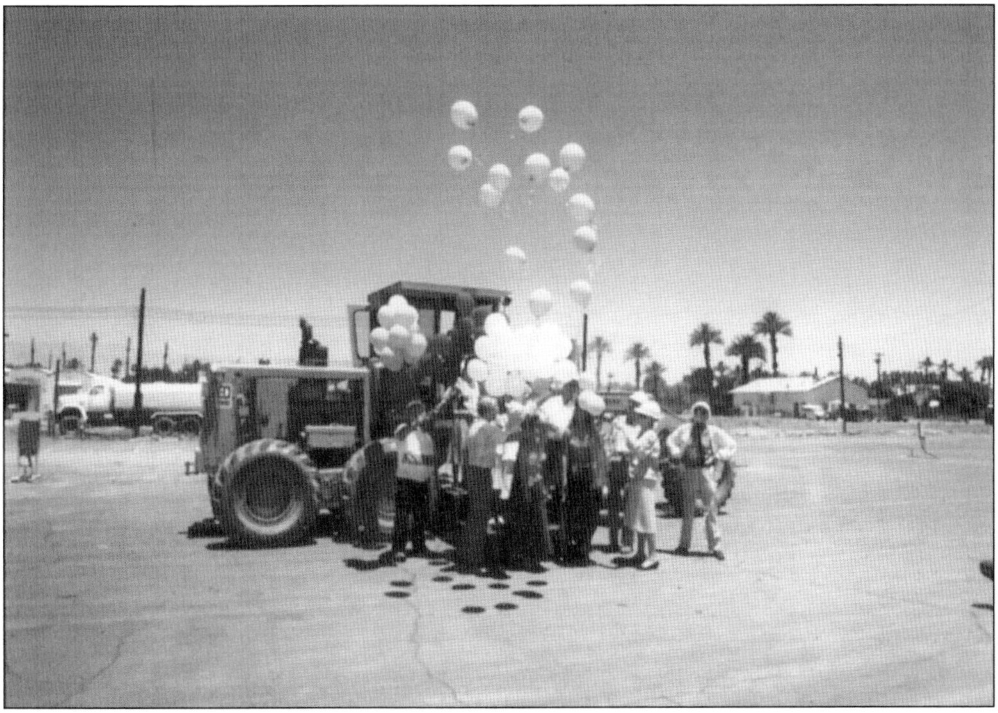

The Fullenwider Building, named after Robert M.C. Fullenwider, who managed the fair from 1947 to 1964, replaced the Palace of Commerce in 1990. Fullenwider had overseen the development of the fairgrounds with their Arabian theme, cheered the development of the pageant, and introduced the camel and ostrich races. Above, balloons commemorate the ground-breaking of the Fullenwider Building. Right, the design features subtler nods to the Greater Middle East than earlier architecture such as the stage and Palace of Commerce. (Above, courtesy of the Coachella Valley History Museum; right, photograph by the author.)

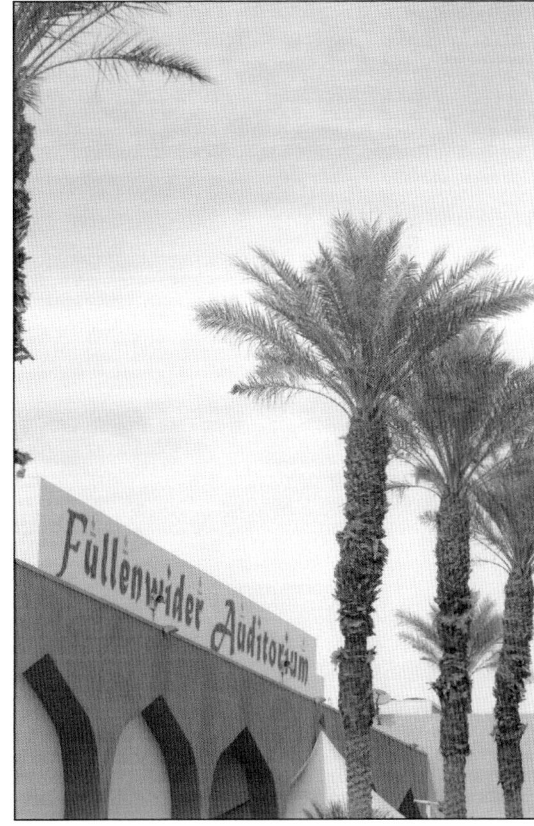

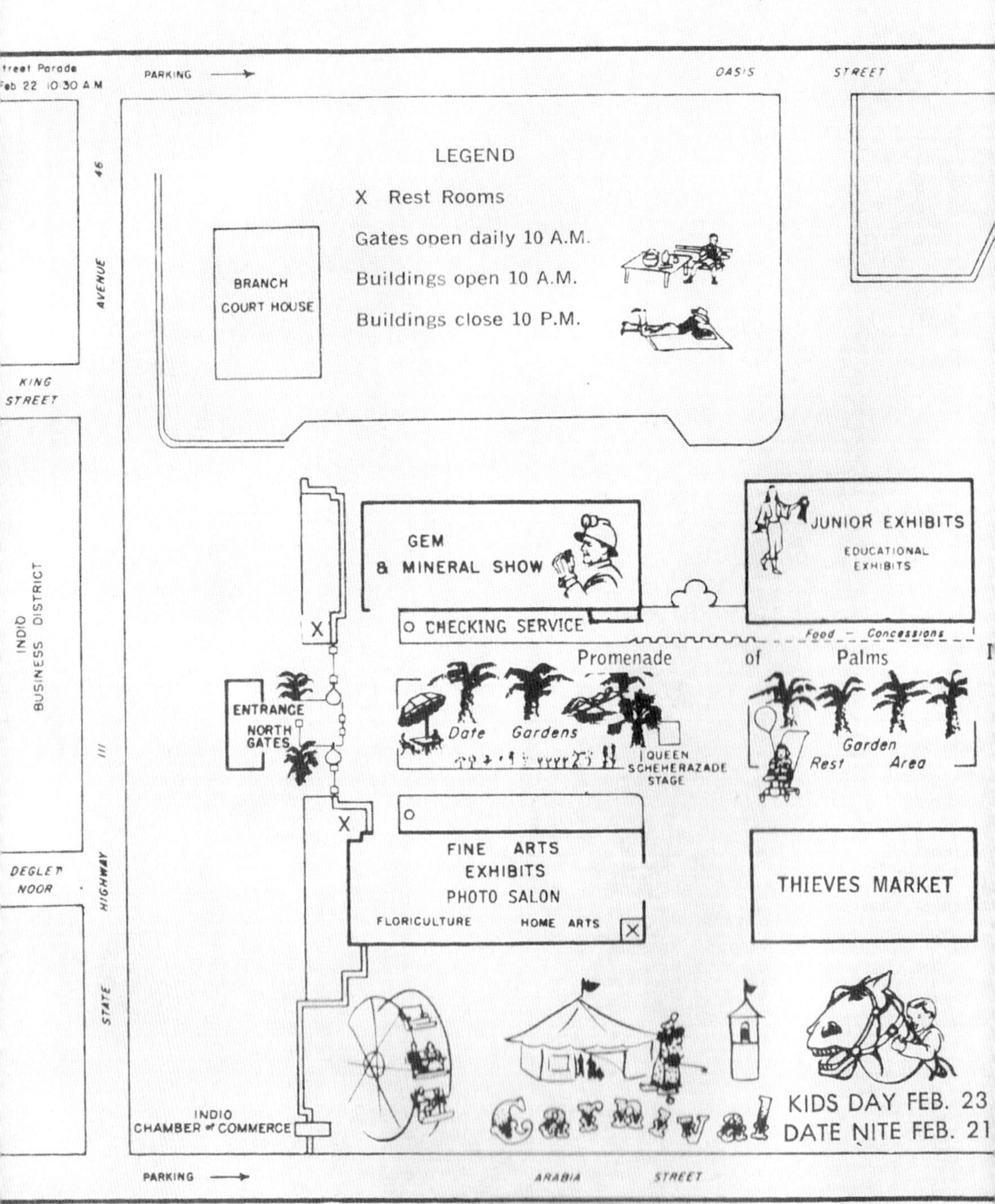

This map from the 1969 program reveals a layout similar to today's date festival. While the fine-art and photography exhibits have replaced the cafeteria in the Caravansary, most other exhibits remain in these locations, including the gem and mineral show and the junior exhibits. Though

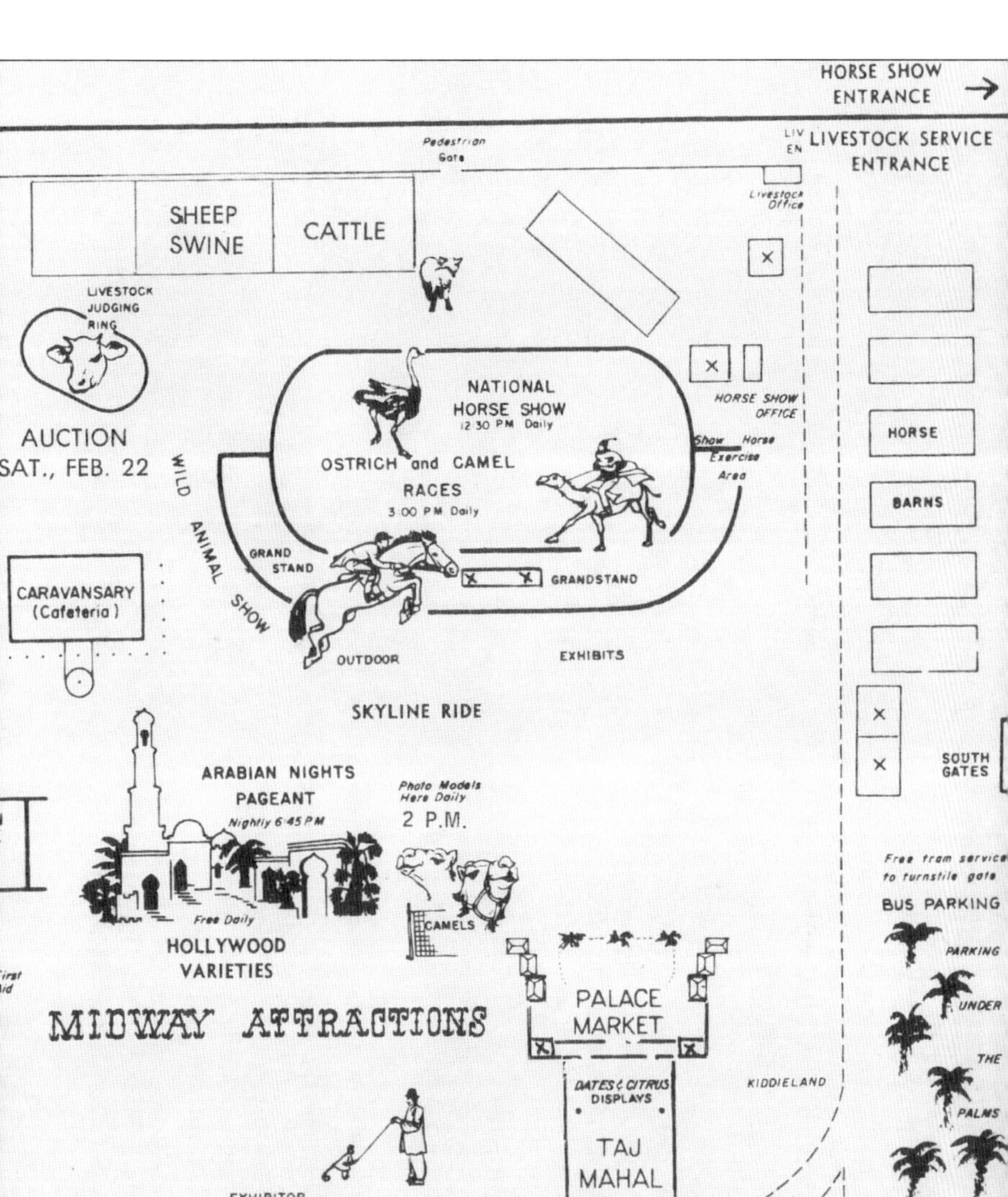

the horse barns have been demolished, the animal buildings occupy the same section. The south entrance has been pushed back to make way for additional carnival rides and vendors, and the parking under the palms was removed long ago. (Author's collection.)

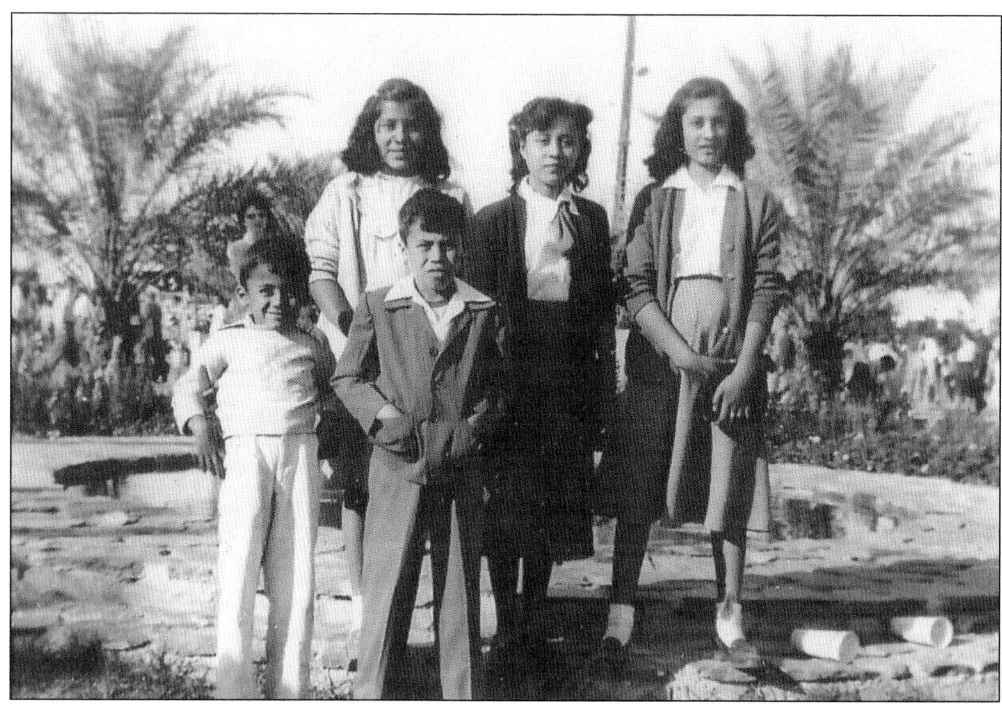

Fountains imbued the festival with a sense of exotic oases, cool respites in the warm desert, lush with flowing water and palm trees. Near the front of the fairgrounds, a stone fountain surrounded a statue of Queen Scheherazade beginning in 1952. Above, from left to right, posed in front of the fountain and obscuring the statue, are (first row) Eddie and Tony; (second row) Olga, Josie, and Ramona Garcia. Below, from left to right are Joe, Inez, and Aaron Leon standing in front of a Persian-style fountain that rested near the Taj Mahal building. Traditionally, date shops, including the Laflin Date Shop seen on the right, sold their products in this area. (Above, courtesy of Ramona Garcia Luna; below, courtesy of the Leon family.)

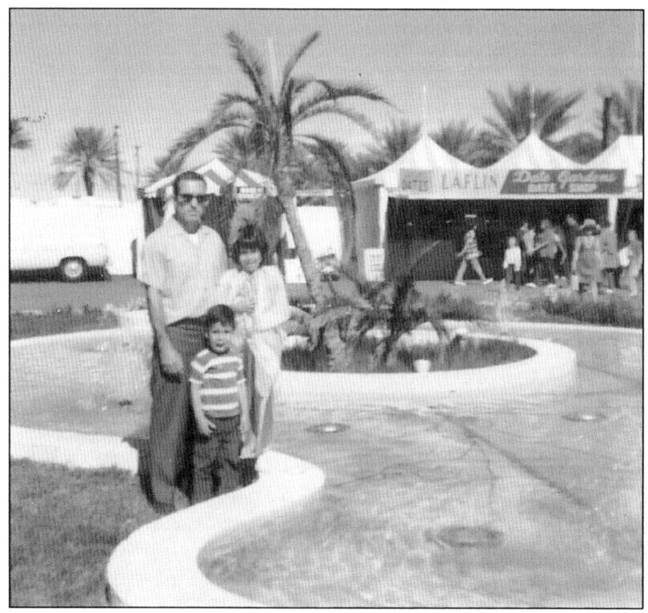

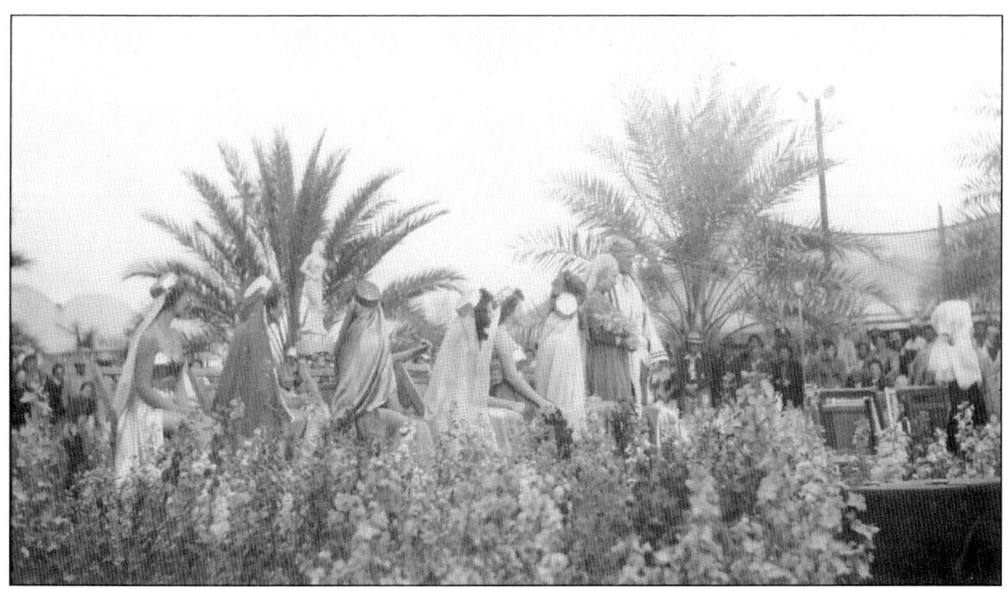

The legend of Queen Scheherazade enveloped the festival. In 1952, she found a place of honor near the front of the fair. There a statue of the mythic storyteller sat in the middle of a flowing fountain, though it was later relocated to the Shalimar Building. Sculpted by Margarita Kerschner, the piece was unveiled to great fanfare only to be stolen from the fairgrounds in 1981. It was returned and eventually found its way to the Coachella Valley History Museum's archives. The unveiling ceremony included the queen's court (above) and the sculptress, standing in the center with flowers next to county supervisor Homer Varner. Kerschner (center right) poses with her husband, Omar Kerschner (far right), who oversaw the gem and mineral building for several years (below). (Both, courtesy of Linda Beal and the History Hunters.)

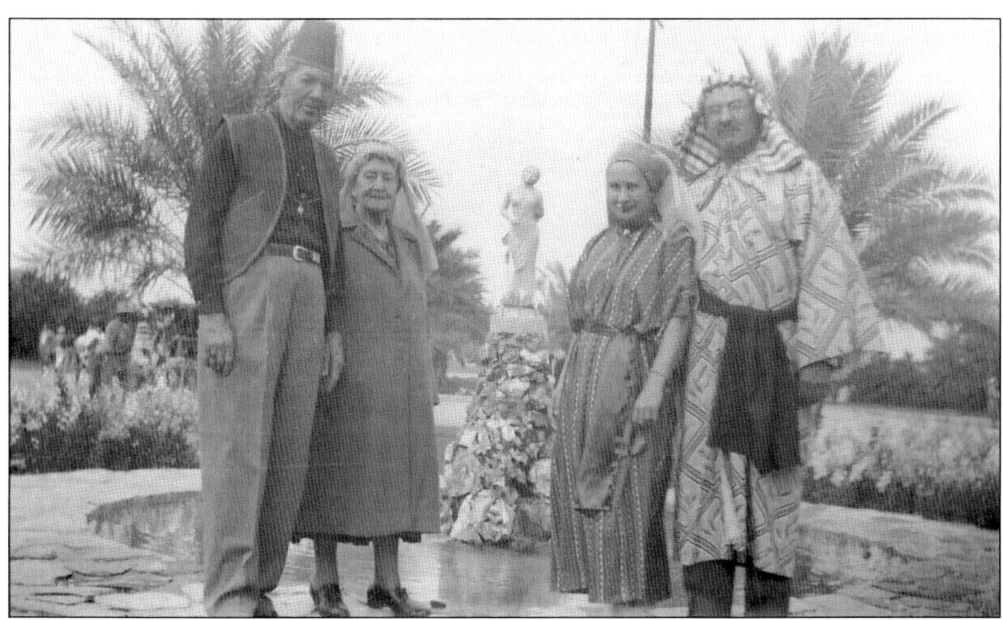

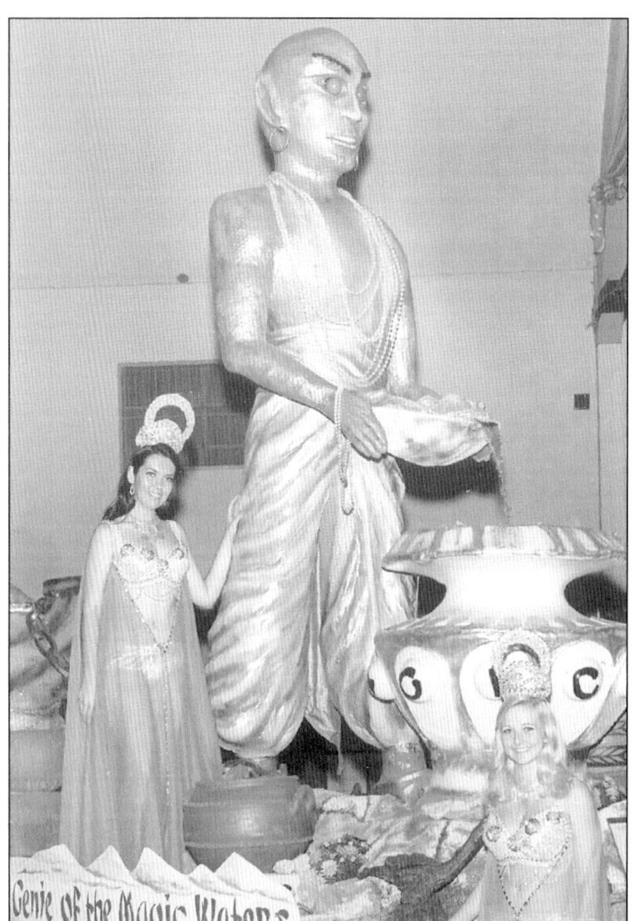

An oversized genie and his magic lamp joined the fairgrounds in the early 1980s, courtesy of sculptor Tony Barone. Even today, they greet fairgoers at the south gate. Left, the members of the queen's court smile with the statue when it was displayed indoors. Below, Barone completes the roadside attraction at his home in Palm Springs. (Both, author's collection.)

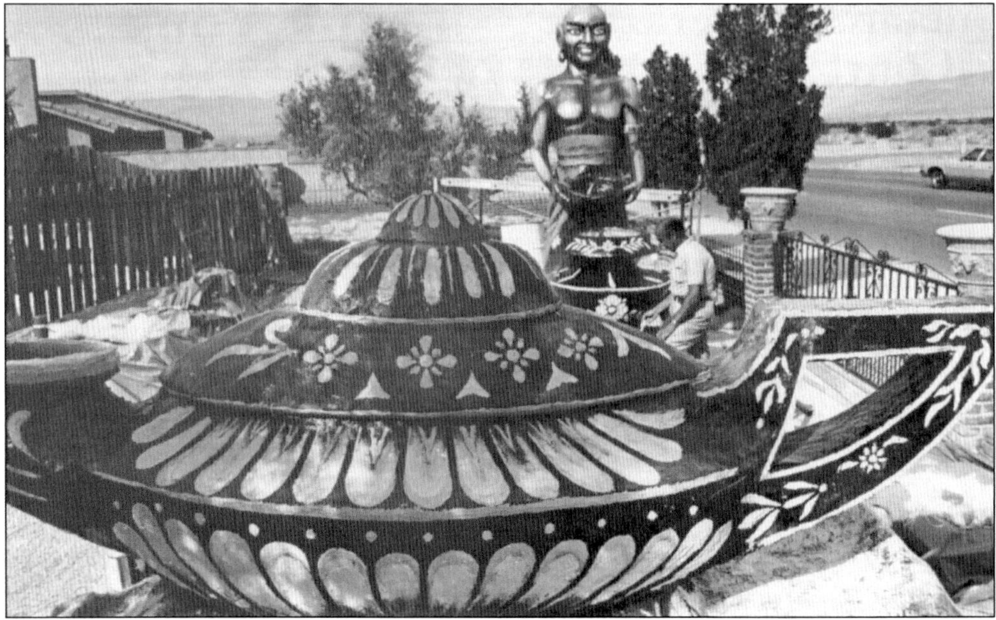

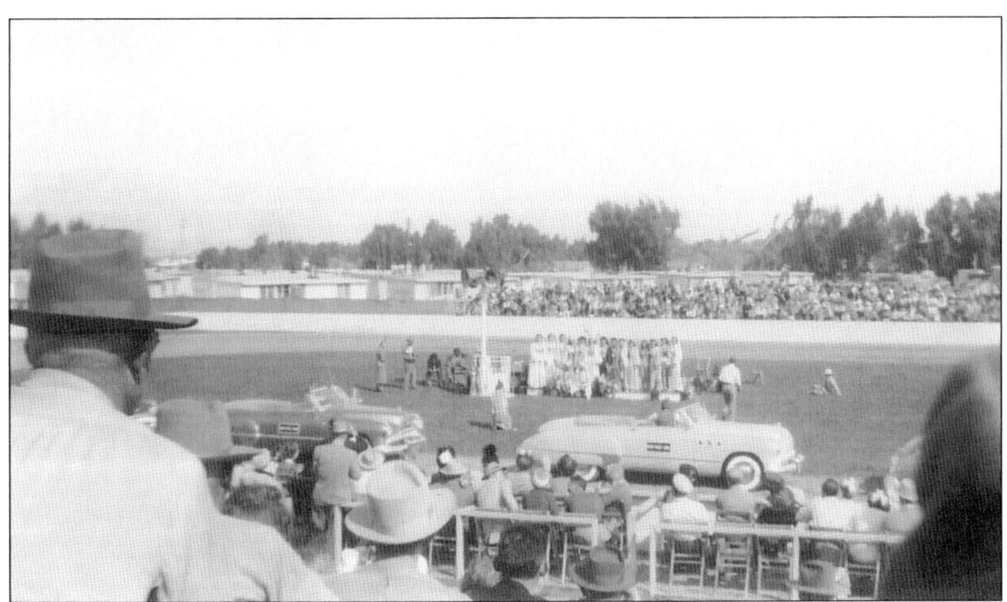

The horse show and rodeo drew huge audiences at the fair, but in the early years, attendance required a separate entrance fee. This 1949 photograph shows the various seats available for purchase, including the box seats toward the bottom, which ran for triple the price of general admission. (Courtesy of Linda Beal and the History Hunters.)

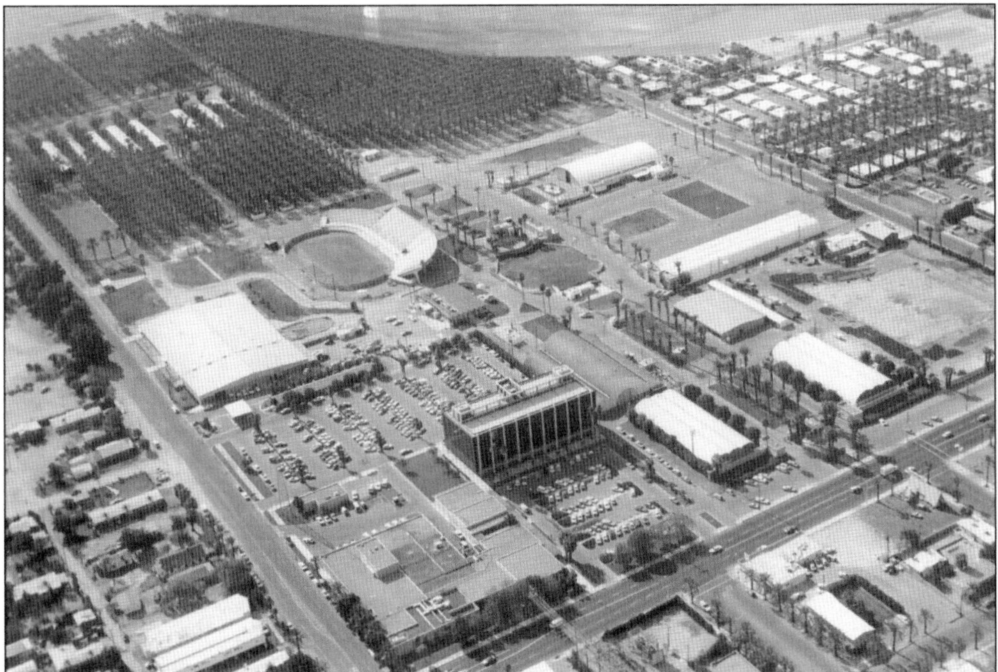

The fairgrounds expanded into the nearby date and citrus groves, adding a large grandstand arena and livestock building, seen here in the middle left of the photograph. The arena hosts camel races, monster-truck derbies, and headliner musical events. It seats over 3,100 people. The lack of the Shalimar Sports Center, where horse-race wagering occurs, indicates this photograph was taken prior to 1975, when the building was completed.

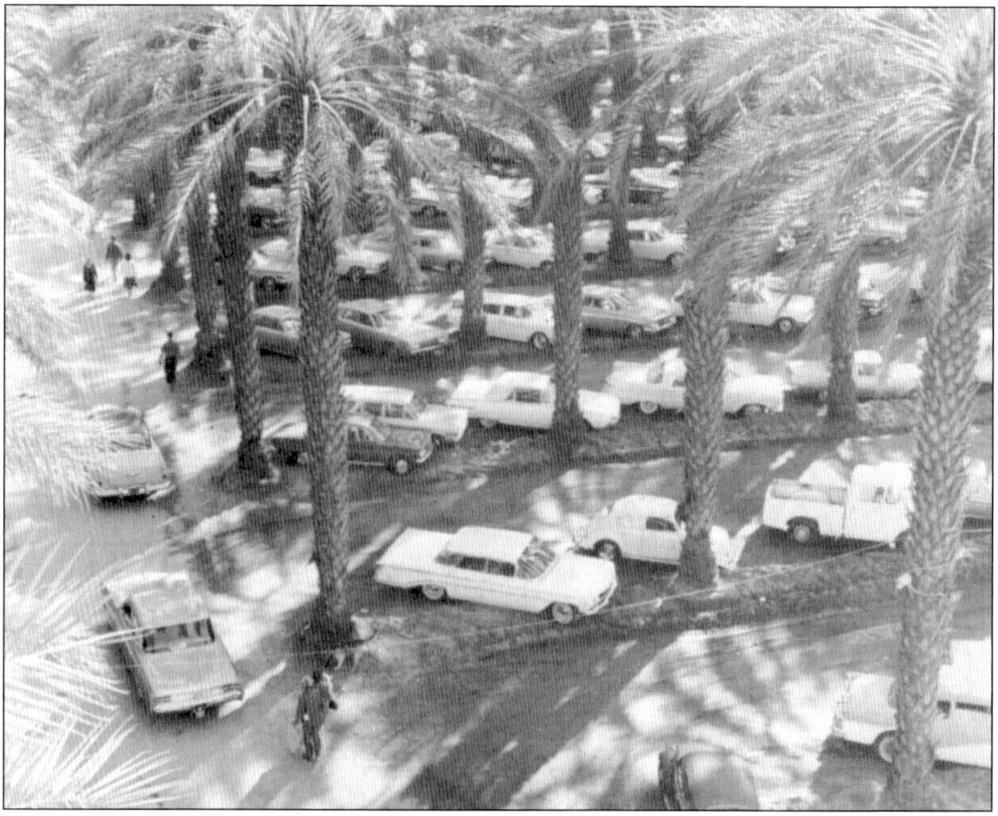

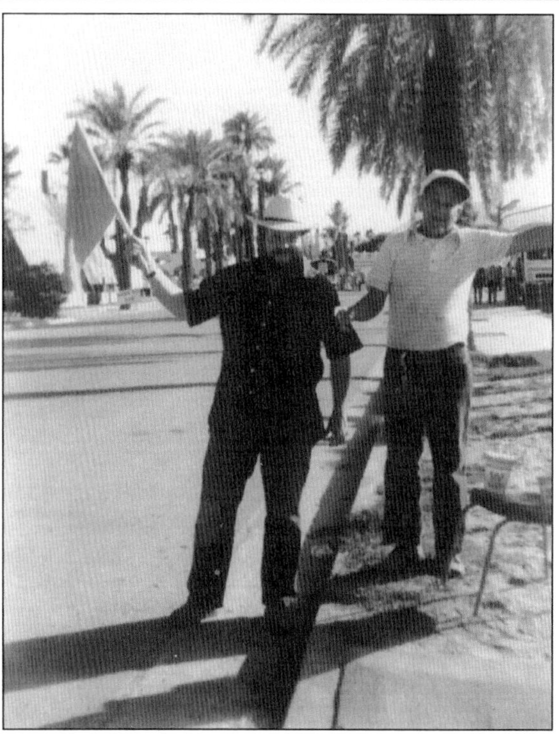

Parking has long been an issue at the festival, especially given so many out-of-town visitors. In 1966, the fair claimed to have the only date-palm parking lot in the nation. Festival neighbors often parked cars on their property to earn extra income during the fair. Rudy Villegas (right) and his son-in-law Jerry Gabriel flag cars into their parking lot, on Las Palmas Court, once home to many prominent Mexican American families in the area. (Above, author's collection; left, courtesy of the Villegas and Gabriel families.)

Landscaping played a vital role in the aesthetics of the fair. Little Inez Leon Esquivel poses in a field of flowers maintained by her grandfather Mike Cabral, the building and grounds foreman at the festival. In addition to paid staff, the festival occasionally put inmates from the nearby Riverside County Jail to work on the fairgrounds—until one escaped. (Courtesy of the Leon family.)

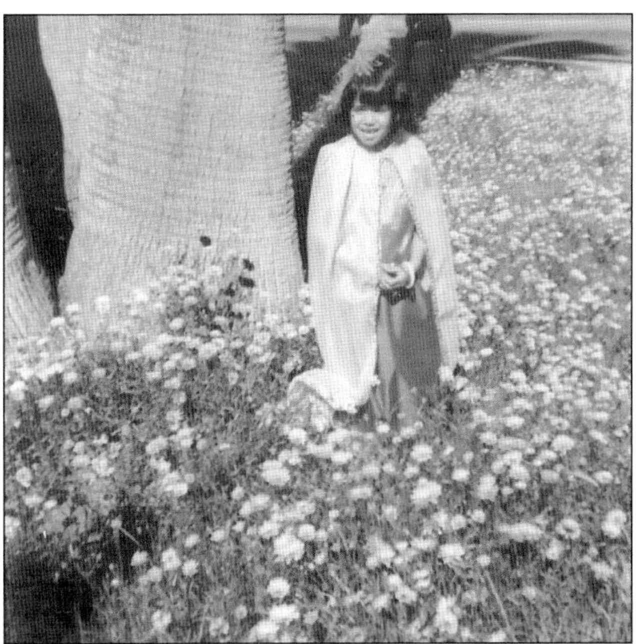

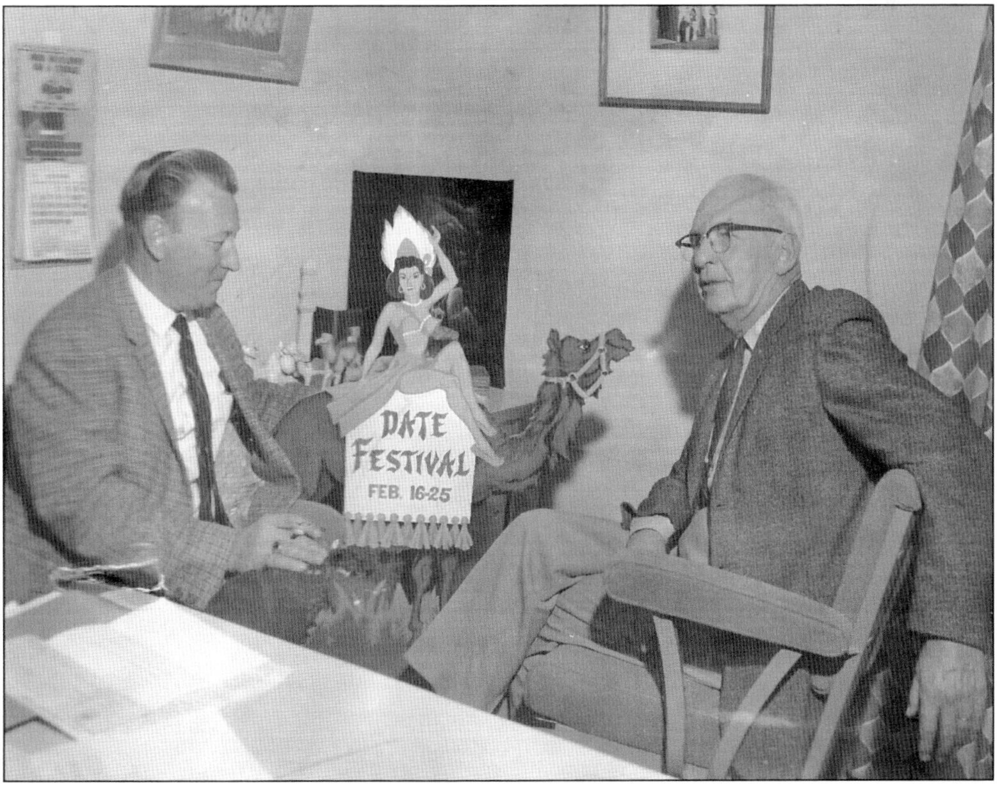

Managers Robert Fullenwider (right) and Charles Wameling review a mockup for a large sign to be displayed outside the fairgrounds. Featuring the two icons of the festival (the camel and Queen Scheherazade), the sign's dates were changed each year. (Courtesy of Linda Beal and the History Hunters.)

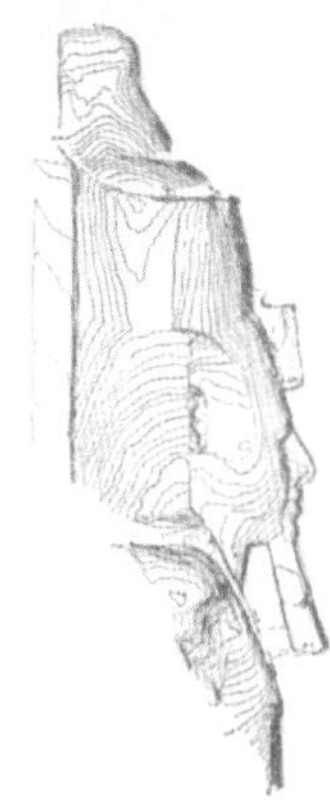

Save the temples
on the Nile

UNITED STATES NATIONAL COMMITTEE FOR THE PRESERVATION OF
THE NUBIAN MONUMENTS

ORIENTAL INSTITUTE
THE UNIVERSITY OF CHICAGO
CHICAGO 37 • ILLINOIS

February 3, 1965

Mr. Arthur L. Wood
82-646 Miles Avenue
Indio, California

Dear Mr. Wood:

It is quite true that the temple of Derr is one of those which has been dismantled and now lies in boxes at the First Cataract for ultimate rebuilding somewhere. I have not been informed whether this rescue is complete and includes the rear rooms which were cut into the solid rock, or whether it has only the front rooms which were built up with masonry blocks. I assume that it was complete.

The city of Indio will certainly have to do something very large, very soon if it is to be in the competition for an Egyptian temple. It is entirely possible that the allocation of monuments and antiquities may be assigned this spring in terms of very clear accomplishment.

Sincerely yours,

John A. Wilson
Executive Secretary

When Egypt set out to build the Aswan Dam, an international movement arose to save the antiquities threatened with submersion. Countries that contributed substantial economic support to relocate the most important temples to higher ground received priority for archeological expeditions and half of the antiquities unearthed. Incredibly, they also pledged some of the smaller temples that otherwise would have been submerged. By 1960, after nationwide coverage of this issue, residents in the Coachella Valley put in a bid to relocate the Temple Derr, hoping to place it at the date festival fairgrounds or the nearby desert. With backing from local and county governments and eventually the State of California, the $1.3-million bid was accepted. Indio failed to raise enough money for the project, especially after the cost increased. Correspondence from the project reflects Indio's interest in the temple.

Three

Royal Reigns
The Queen Scheherazade and Arabian Nights Pageants

The original inspiration for the fair's theme stemmed from the widely read *One Thousand and One Arabian Nights*. Nowhere was this more evident than in the fair's famed pageants: the Queen Scheherazade Beauty Pageant and the theatrical *Arabian Nights* musical pageant. The collection of tales arose in the medieval Middle East, translated for Western audiences as early as 1706 CE. While different translations included different stories, they all centered on the main tale of Queen Scheherazade. After executing his adulterous spouse, King Shahryar marries a series of women, each of whom he beheads after one night to prevent them from cheating. Scheherazade eventually offers herself as his new bride. On the night of the wedding, she begins to tell the king an intriguing tale, stopping before the story's resolution. Curious, the king has little choice but to let her live. The next night she finishes the story only to start another one, continuing the process for 1,001 nights.

Known for both her beauty and her cleverness, Scheherazade proved an interesting inspiration for the beauty pageant. Though there were queens in earlier festivals, the title became Queen Scheherazade in 1948. Fair manager Robert Fullenwider's daughter Florobel is credited with suggesting the association. Since then, hundreds of young women have competed for the title. Winners represent the county at official events and roam the fairgrounds in the iconic harem-girl outfits.

Proposed by Louise Dardenelle in late 1947, the first pageant debuted on the newly constructed stage in 1948. Every year, pageant directors chose a different tale from *One Thousand and One Arabian Nights* to reenact. Because the tales Scheherazade wove traversed North Africa, the Middle East, and Asia, they proved exceptional fodder for varied theatrical performances. At first, locals supplied all the talent needed, both on stage and behind the scenes, though professional help was hired in later fairs. Staged annually for over 65 years, it is California's second-oldest outdoor play, outrun only by the *Ramona* pageant in nearby Hemet.

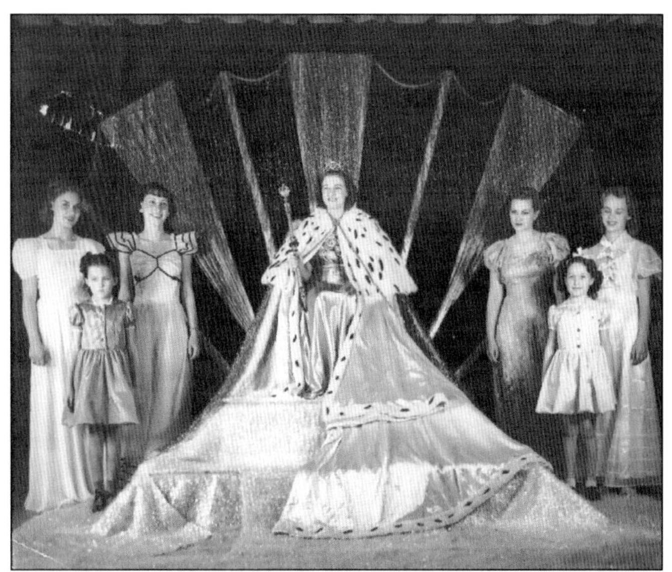

Doris Creveston, representing the Coachella Valley, was crowned queen of the 1940 Riverside County Fair. The Women's Club, Exchange Club, Coachella Valley Lions, and Catholic Youths' Organization sponsored individuals for a local contest. Creveston, supported by the Indio Chamber of Commerce, competed against young women from other towns in the county. In these early years, the public voted via ballots obtained through purchases with local merchants.

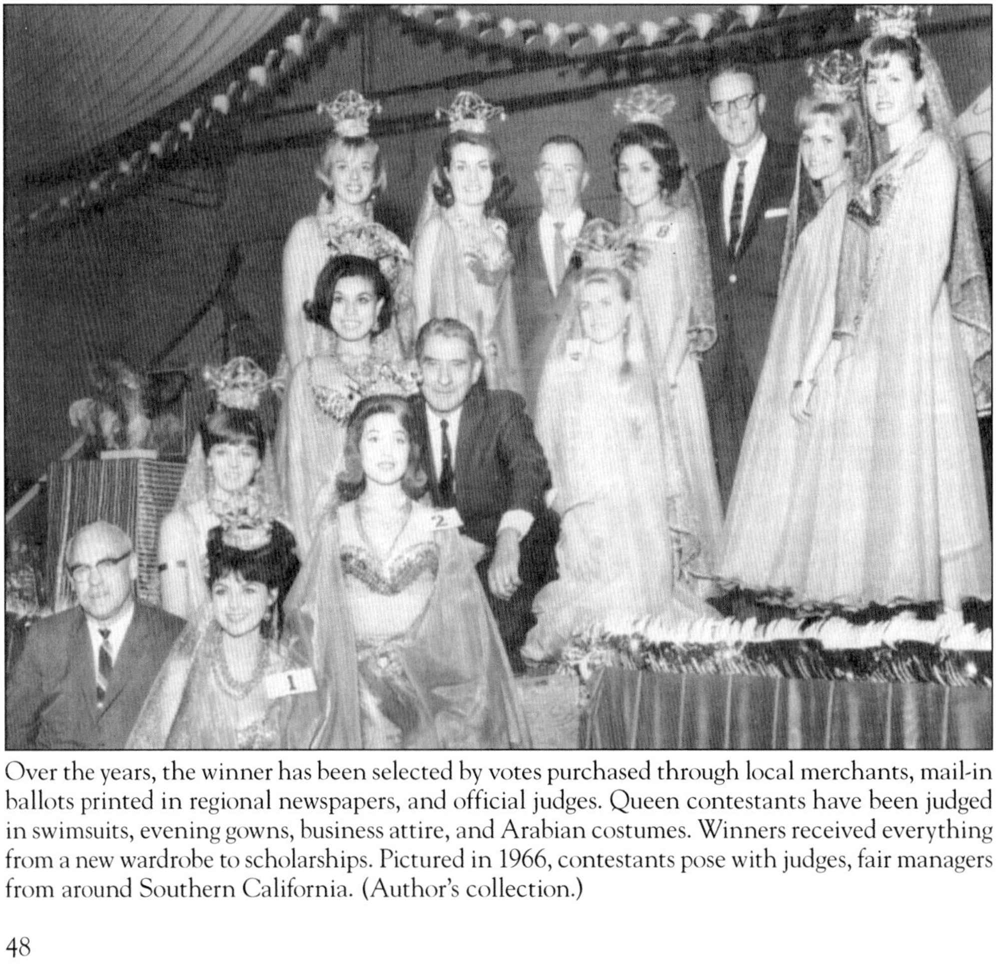

Over the years, the winner has been selected by votes purchased through local merchants, mail-in ballots printed in regional newspapers, and official judges. Queen contestants have been judged in swimsuits, evening gowns, business attire, and Arabian costumes. Winners received everything from a new wardrobe to scholarships. Pictured in 1966, contestants pose with judges, fair managers from around Southern California. (Author's collection.)

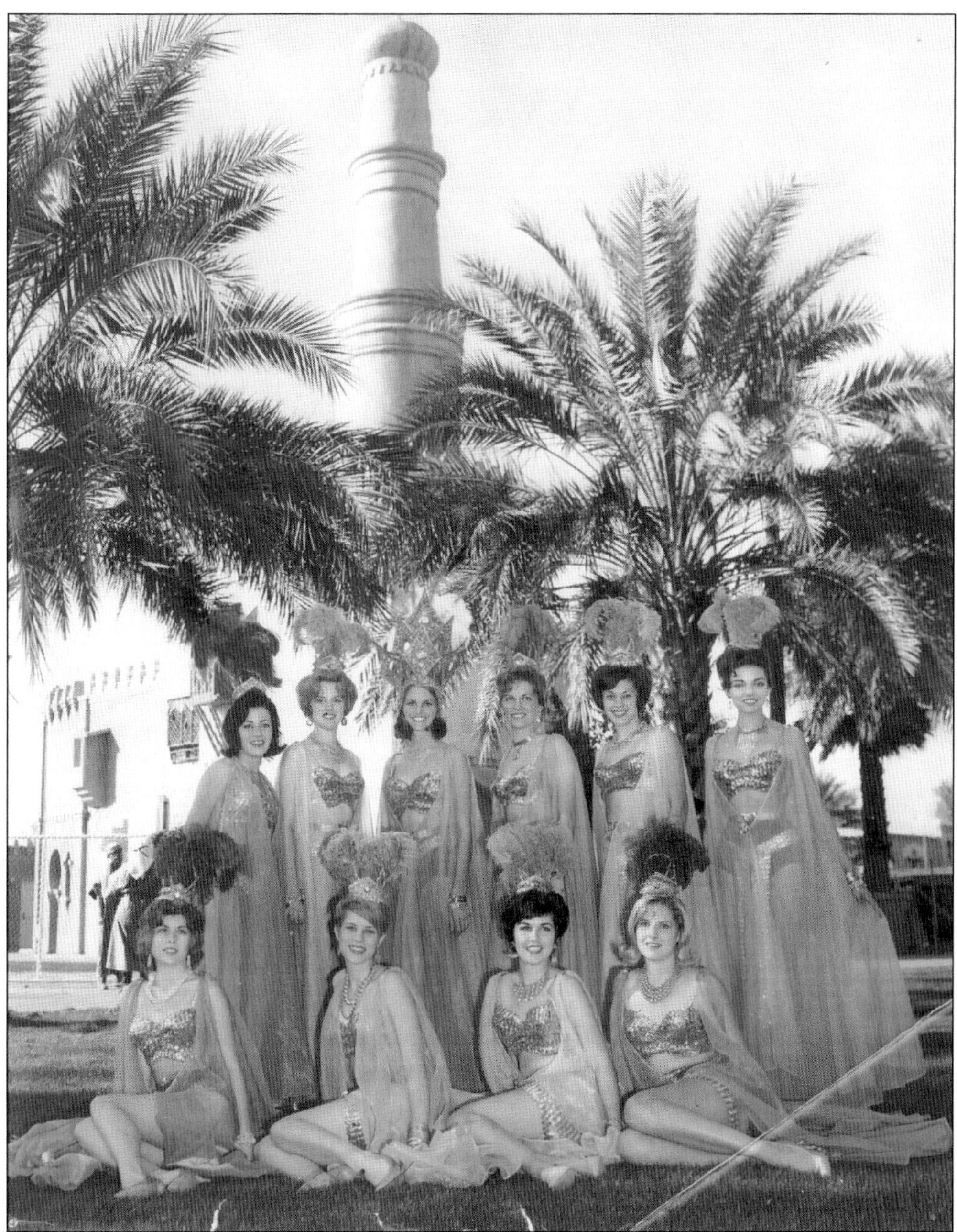

The number of princesses on the court varied. Often, each large city in Riverside County—Riverside, Hemet, Banning, San Jacinto, Elsinore, or Palm Springs—sent a representative to compete against the Coachella Valley's choice. By 1977, the court dwindled down to just five members, chosen from finalists throughout the region. By 1983, the current model was in place. The beauty pageant was open to the entire county, but just three contestants went on to become Queen Scheherazade or Princesses Dunyazade and Jasmine. In *One Thousand and One Arabian Nights*, Dunyazade is Scheherazade's little sister, who prompts her storytelling in an effort to escape beheading. The 1964 court, pictured here, had 10 members, typical for the 1960s. (Photograph by Roy Gillman.)

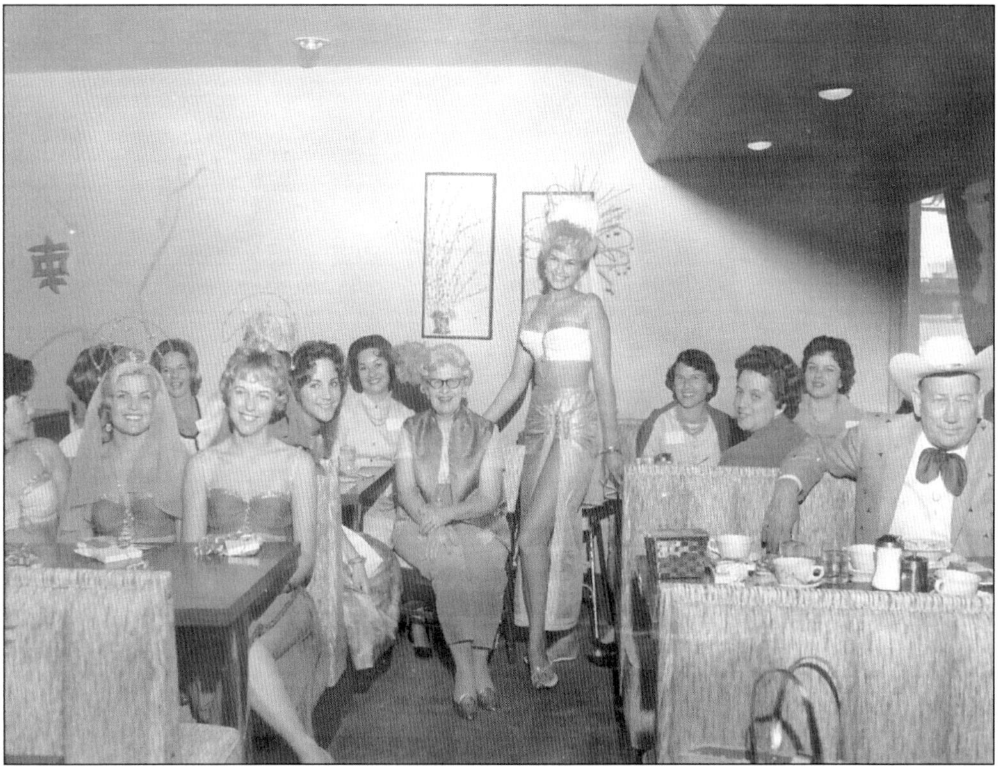

Local businesses hosted meals for the women at well-loved restaurants such as Rancho Carrillo, Frances's Café, El Morocco, the Brite Spot Café, Sans Souci, Ciro's, the Plaza Hotel, and the La Quinta Hotel. Clubs also fed the girls, occasionally in the home of one of their members or in their lodges. (Courtesy of Linda Beal and the History Hunters.)

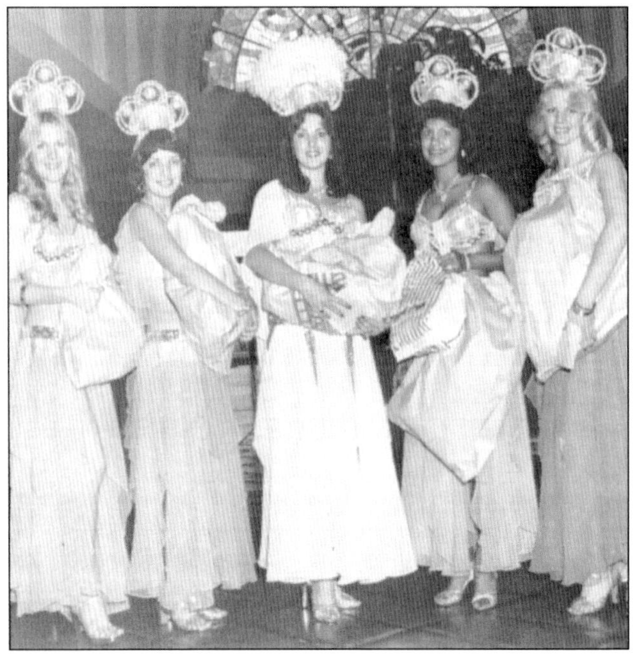

As representatives of the National Date Festival and Riverside County, the queen and her court made public appearances at special events and businesses around the area. Here, the court poses at the Indio Fashion Mall, its iconic stained-glass window visible behind their heads. (Author's collection.)

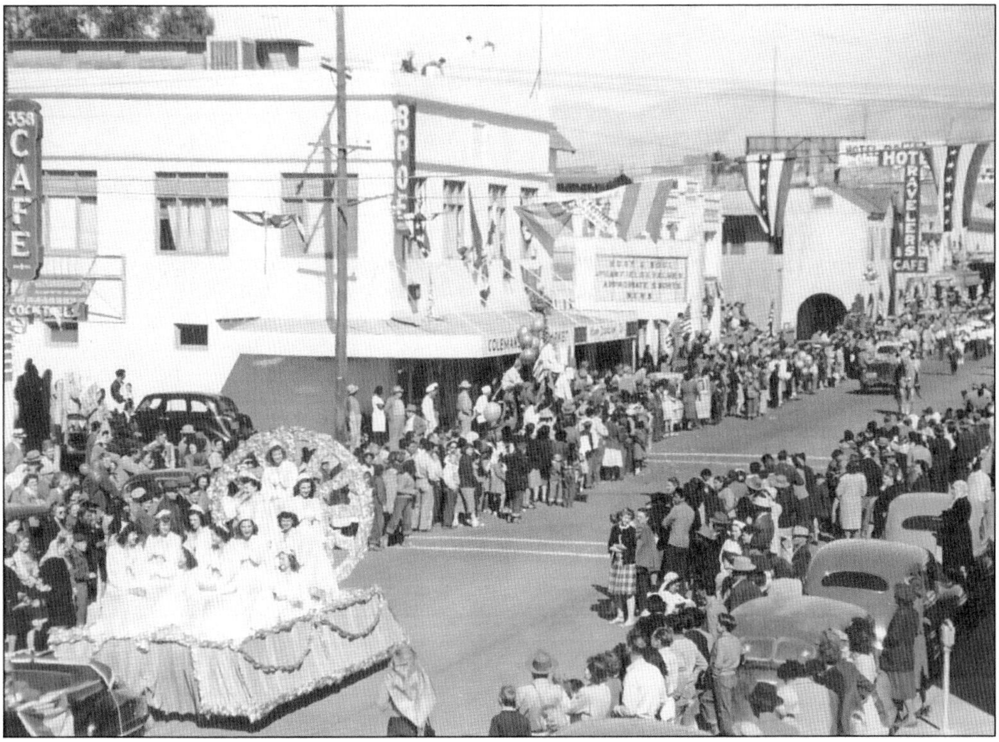

Part of the court's responsibilities included riding in the parade. Each year, spectators looked forward to the queen's float. This one featured the 1948 court with matching harem outfits and its queen, Janet Anderson Nicks. During early festivals, local social clubs or fraternal organizations created these floats. (Courtesy of Janet Anderson Nicks.)

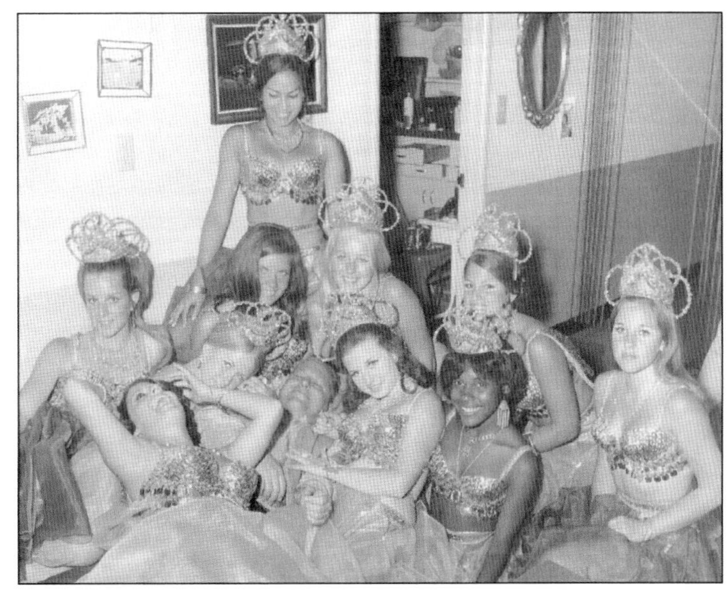

The majority of queen contestants were white, though some Mexican American women served as princesses and eventually as queens, especially as the Mexican and Mexican American population in the Coachella Valley grew. Few African American women served on the court. Debbie Ratchford of Moreno Valley was the first black princess, in 1970.

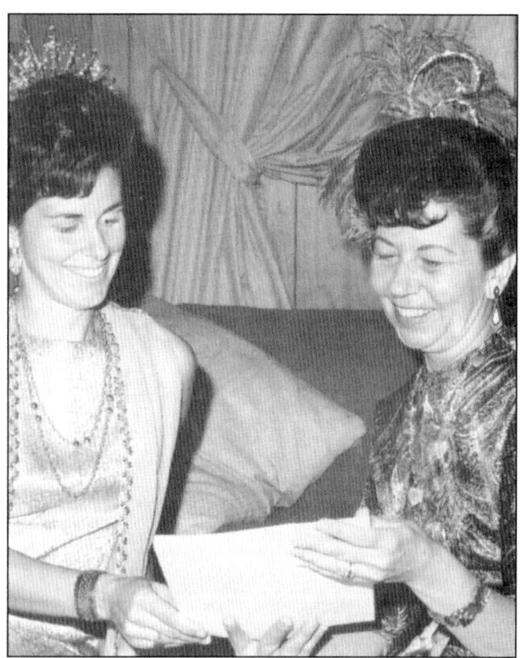

The queen mother chaperoned the young ladies, accompanying them to official events, including television and radio appearances. Originally, the local Women's Club jointly filled this role. Remembered kindly as Mother Ora by contestants, queen supervisor Ora Jared (right) held the position for more than 10 years. Sandra Krause (left), once a court member herself in 1954, replaced Jared in 1970. (Author's collection.)

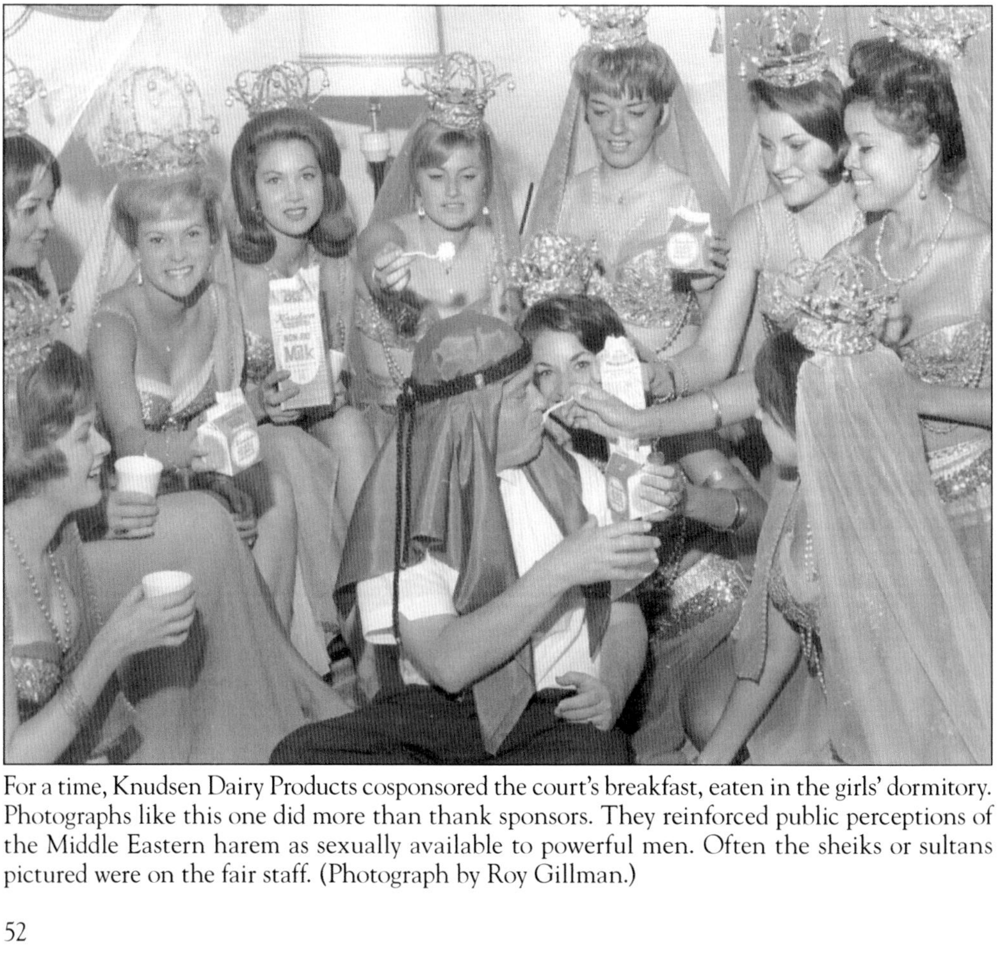

For a time, Knudsen Dairy Products cosponsored the court's breakfast, eaten in the girls' dormitory. Photographs like this one did more than thank sponsors. They reinforced public perceptions of the Middle Eastern harem as sexually available to powerful men. Often the sheiks or sultans pictured were on the fair staff. (Photograph by Roy Gillman.)

Because the court remained at the fair during its run, its members missed several days of school. Since most of the young women lived elsewhere in Riverside County, they stayed together in Indio under the supervision of chaperones. In the 1940s, contestants resided at the Hotel Indio, then the Biltmore Hotel. Eventually, the girls stayed in dormitory-style accommodations near the fairgrounds. (Courtesy of Linda Beal and the History Hunters.)

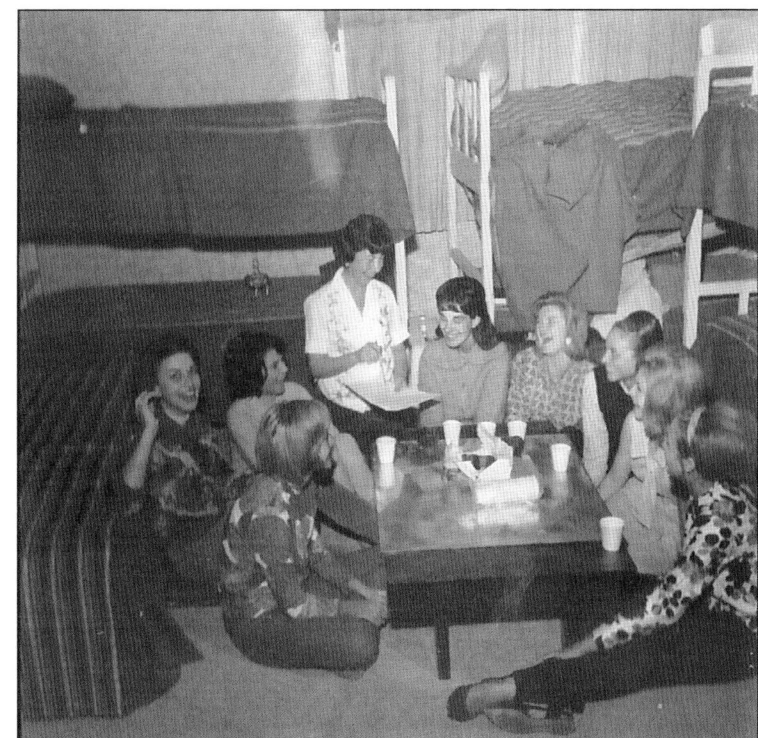

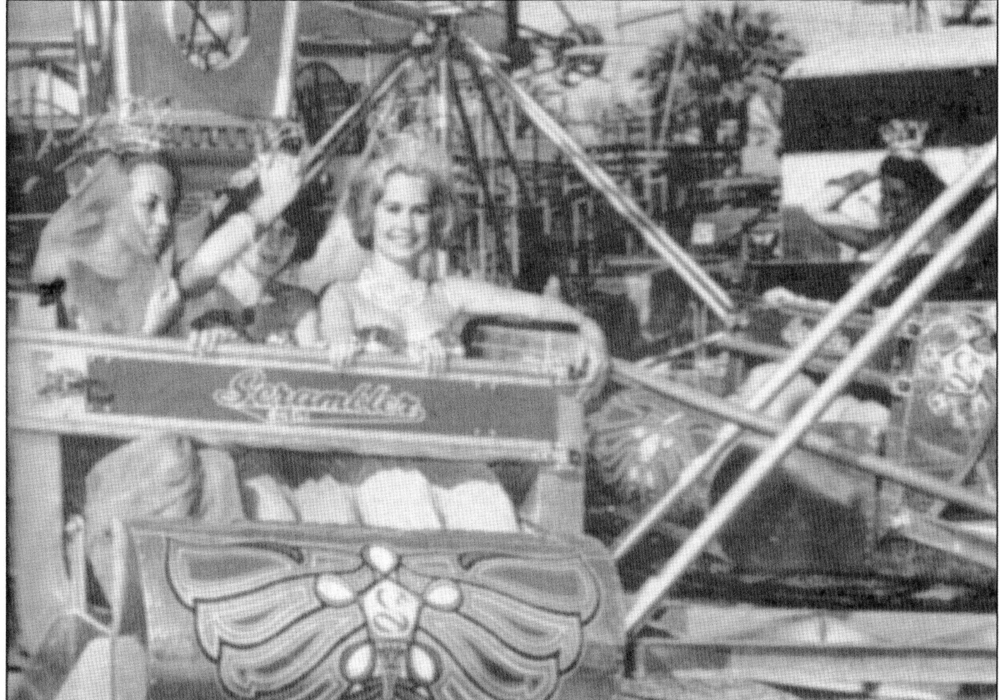

During pre-fair publicity photo shoots, the queen and court check out all the attractions of the fair, even before the public. Members of the court enjoy the scrambler ride, around 1965. (Author's collection.)

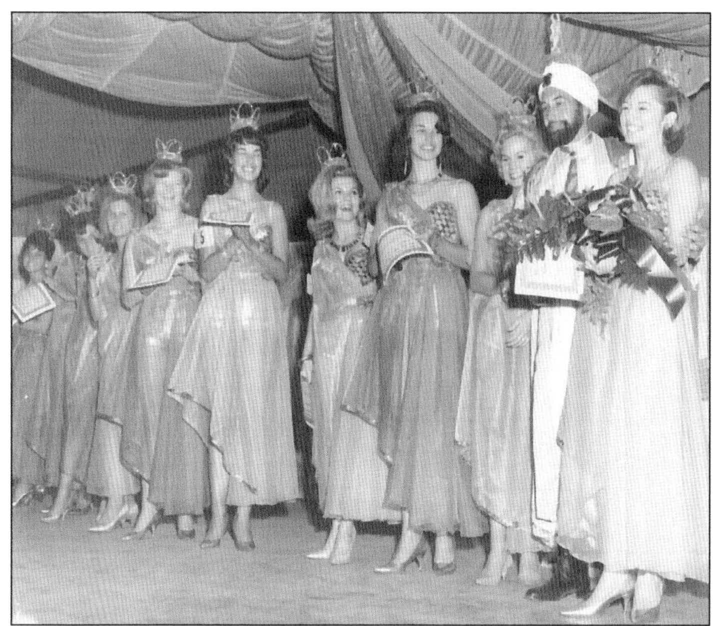

Two queens reigned in 1965. Miss Palm Springs, Betty Hendrickson (far right), won the title. She married the same year, though contest rules stated the winner must remain single. Organizers polled the queen's court, and Indio's Sandra Slocum (fifth from left) was chosen as a replacement. The unprecedented series of events led to the creation of the Princess Dunazade and Princess Jasmeen (Jasmine) titles for first and second runners-up in 1966.

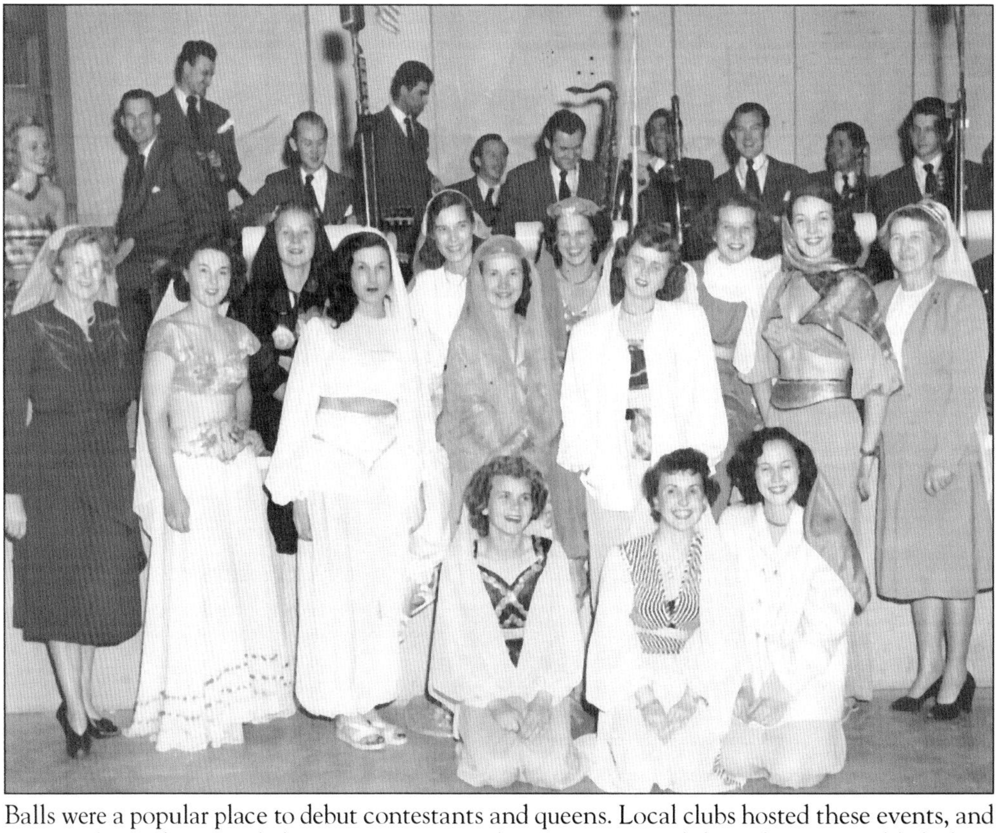

Balls were a popular place to debut contestants and queens. Local clubs hosted these events, and area residents also attended in costume. Here, the 1948 court and their chaperones, likely from the Women's Club, pose in front of a band at one of the dances held in their honor. (Courtesy of Janet Anderson Nicks.)

The queen and court rode in style around the fairgrounds and at the parade. Local car dealerships loaned their best models, both to support the date festival and to advertise their wares. Right, members of the court ride around the arena, likely during the horse show. Below, Imperial Motors, which maintained dealerships in Palm Springs and Indio, served as the official car of Queen Scheherazade. The contestants pose with Imperial Motors' Kay Olsen in front of the Taj Mahal Building. (Right, courtesy of Dean Payne; below, courtesy of Linda Beal and the History Hunters.)

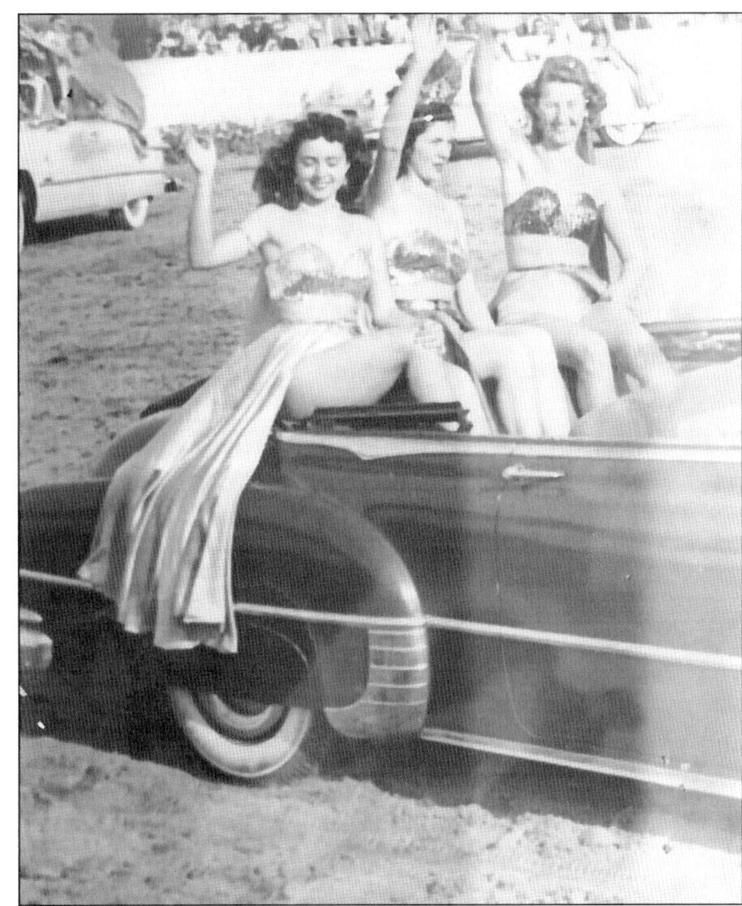

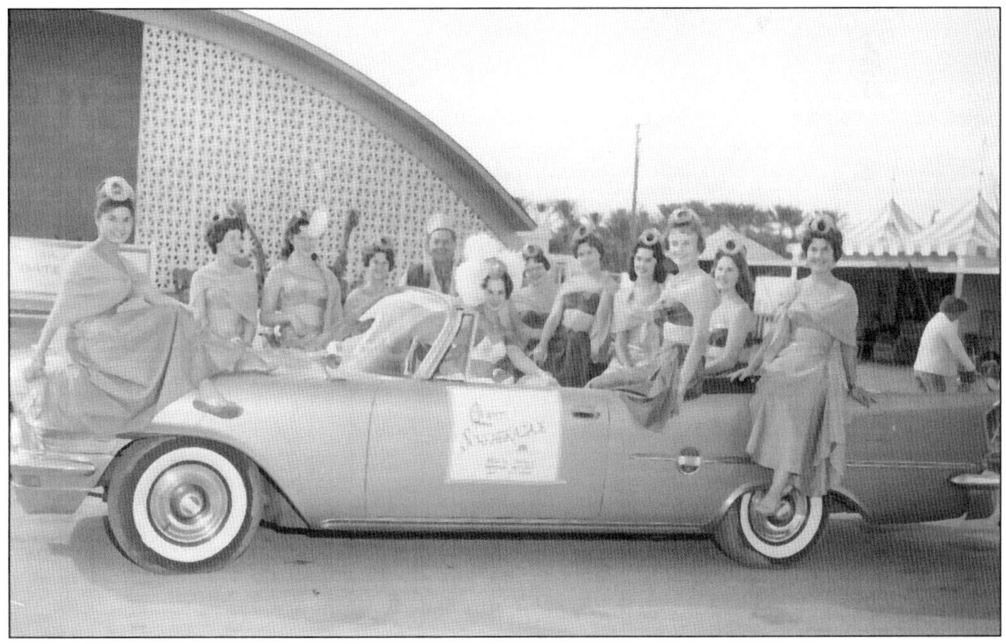

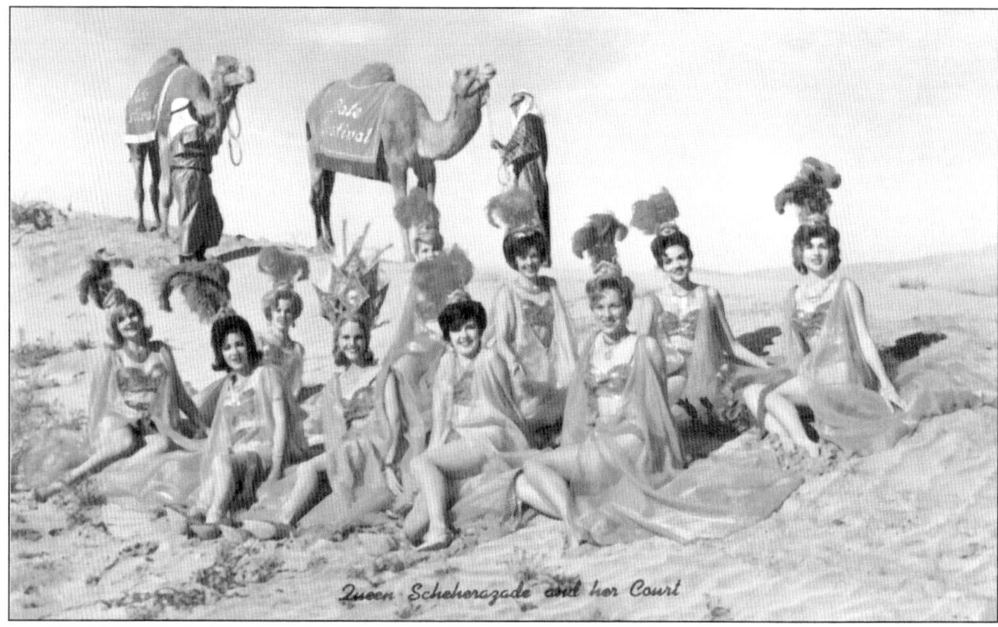

For decades, other fairs and festivals around Southern California sent images of their queens to regional newspapers. These images drew attention not only to the event but to the specific hosting cities as well. Queen Scheherazade sold the fair in her publicity shots but also reminded readers of the region's specialty crop, the date. Most of the fair publicity shots featured harem girls and dates, camels, or the fair's architecture. Sometimes they included the nearby desert landscapes, like these sand dunes from 1964 (above). Images like this were typical for the fair's souvenir program. The court embraces publicist Bill Arballo (below) in his office.

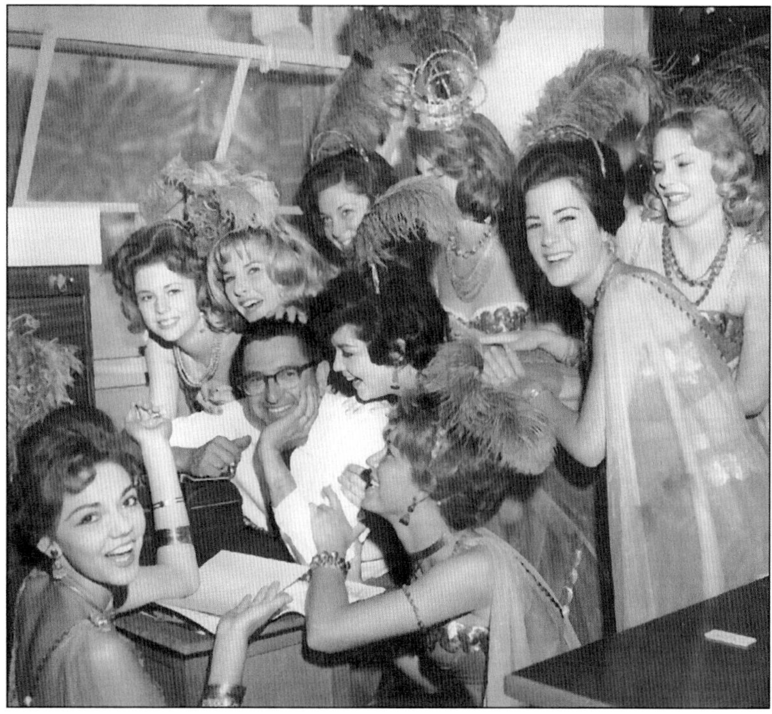

The pageant was truly a community event, with upward of 100 individuals assisting in the production every year. It was also multigenerational; young children sang and danced onstage while older individuals, sometimes participating for more than 30 years, contributed as well. Occasionally, paid actors filled the main roles, but the majority of those involved came from the surrounding community. More than 50 participants can be seen in this undated photograph.

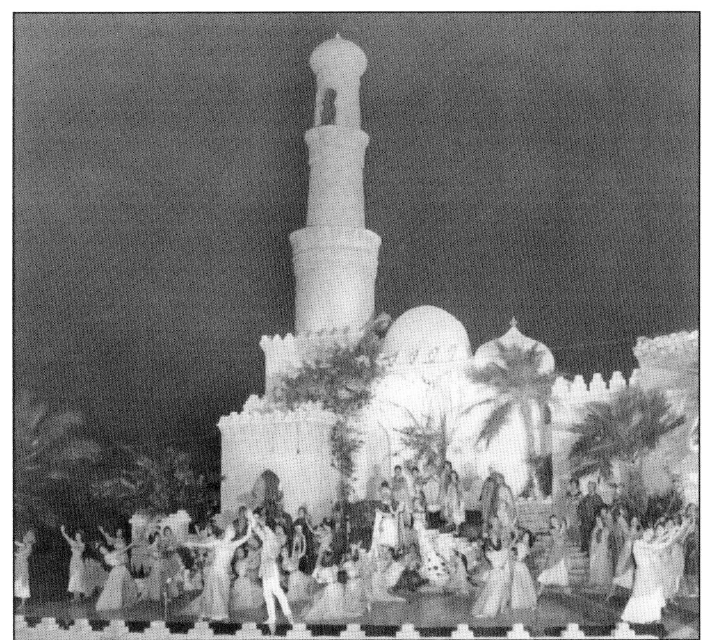

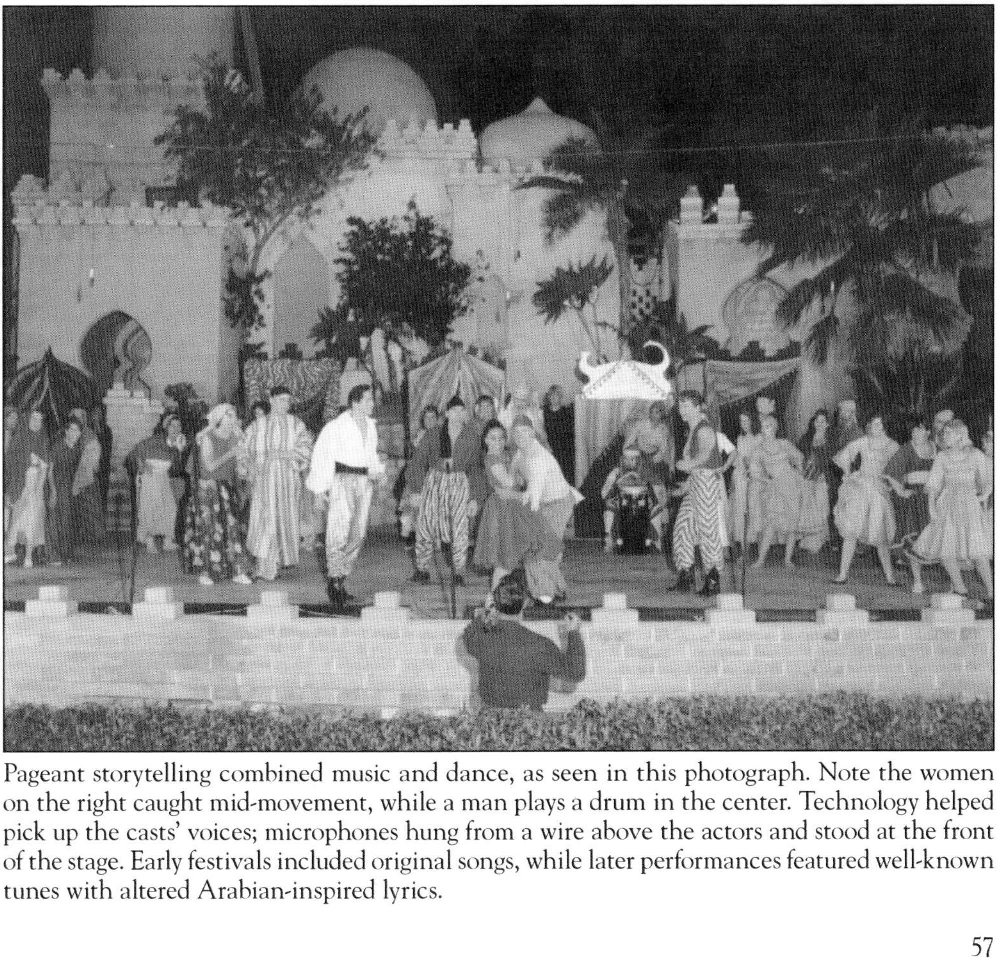

Pageant storytelling combined music and dance, as seen in this photograph. Note the women on the right caught mid-movement, while a man plays a drum in the center. Technology helped pick up the casts' voices; microphones hung from a wire above the actors and stood at the front of the stage. Early festivals included original songs, while later performances featured well-known tunes with altered Arabian-inspired lyrics.

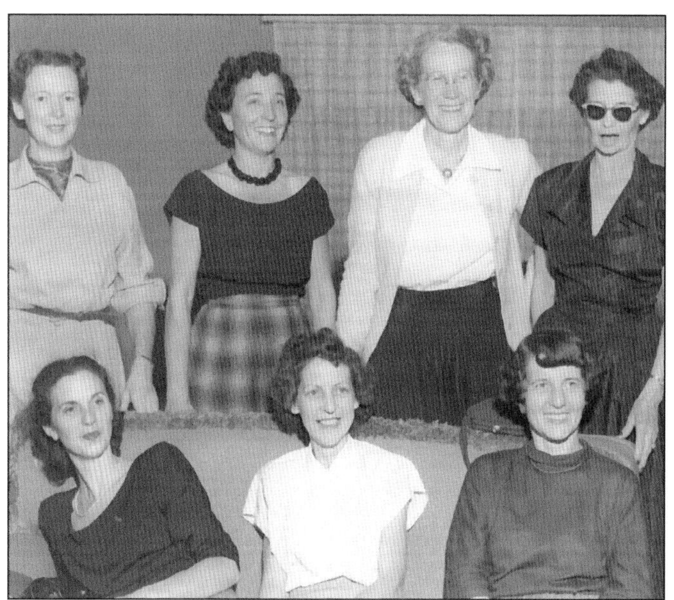

Instrumental to the pageants of the 1950s, pictured standing, from left to right, are production supervisor Dorothy Gavin; Kay King, who supervised musical selections; and director Bertha Young. Young was once so engrossed with the play that she fell from the stage into the orchestra pit, unscathed. Carolyn Tyson (standing, right) headed props and stage design while Joyce Riemer (seated, left) choreographed. Zenda Elliott (seated, center) designed costumes, and Sarah Farrar (seated, right) wrote the scripts.

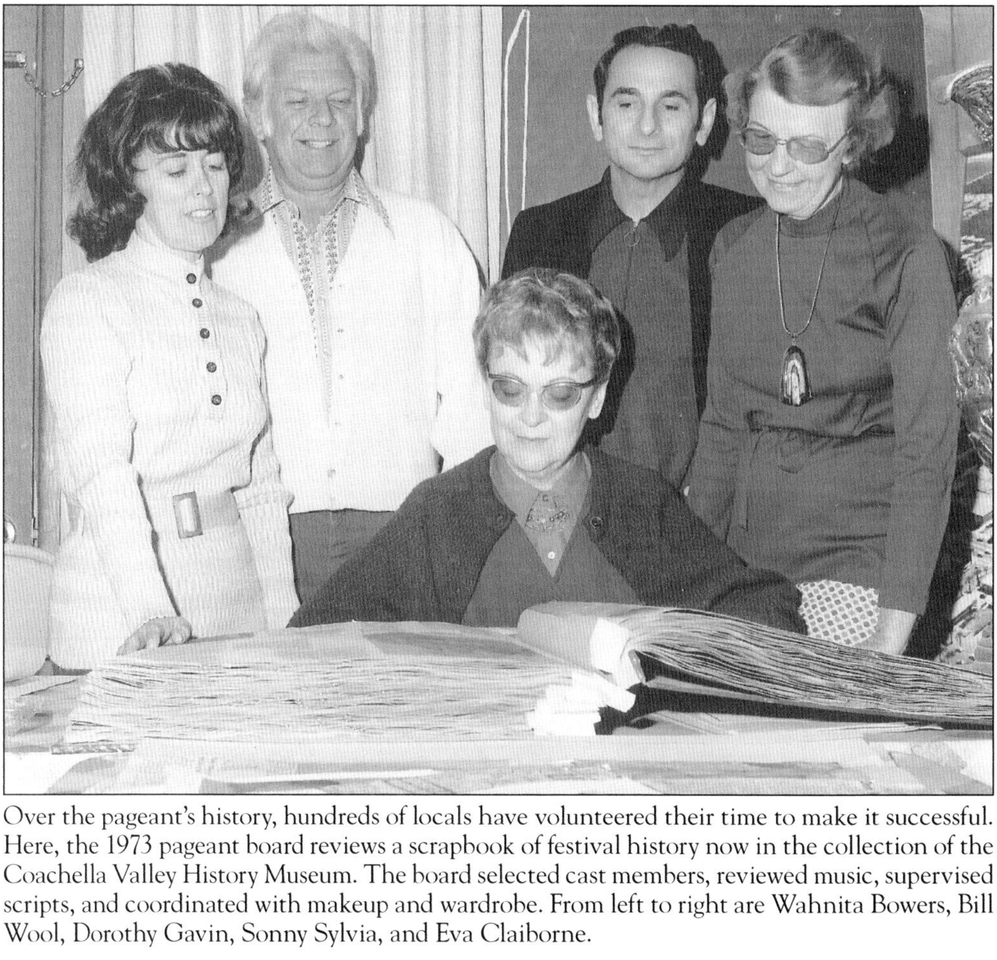

Over the pageant's history, hundreds of locals have volunteered their time to make it successful. Here, the 1973 pageant board reviews a scrapbook of festival history now in the collection of the Coachella Valley History Museum. The board selected cast members, reviewed music, supervised scripts, and coordinated with makeup and wardrobe. From left to right are Wahnita Bowers, Bill Wool, Dorothy Gavin, Sonny Sylvia, and Eva Claiborne.

For more than 30 years, 1952–1983, Sarah Farrar, pictured here, wrote the script for the pageant. While others such as Sonny Sylvia (center) provided the story, Farrar adapted the work, drawing inspiration from Richard Burton's 17-volume edition of the *Arabian Nights*. Doug Caplette (right), a well known local radio personality, served as the narrator before passing on the torch in 1976.

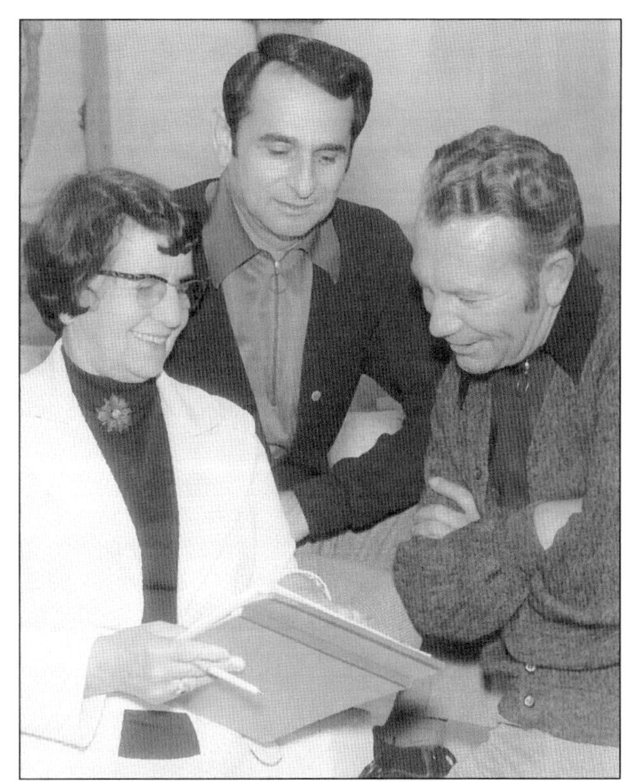

Active since the first 1948 pageant, G.L. Christiansen is remembered for playing the role of the caliph (center) for more than 25 years. When not onstage, Christiansen worked as the deputy county clerk of Indio. Notice his cheetah costar. Animals, including camels, elephants, and horses, were frequent guests onstage, while cheetahs and tigers delighted the audience as well.

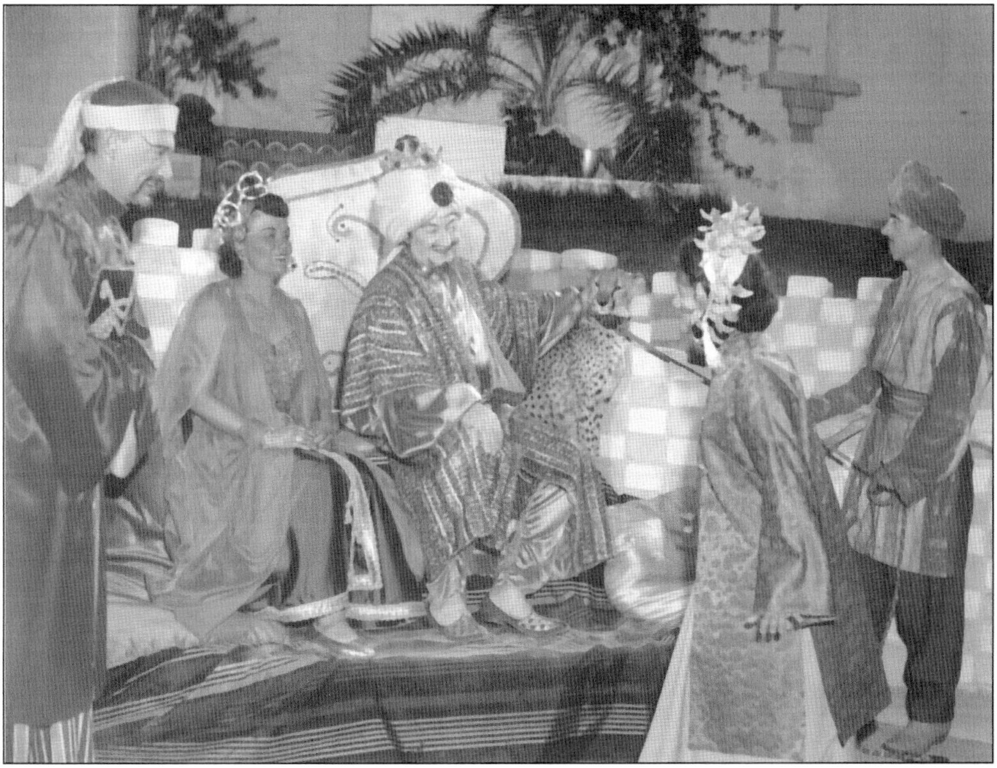

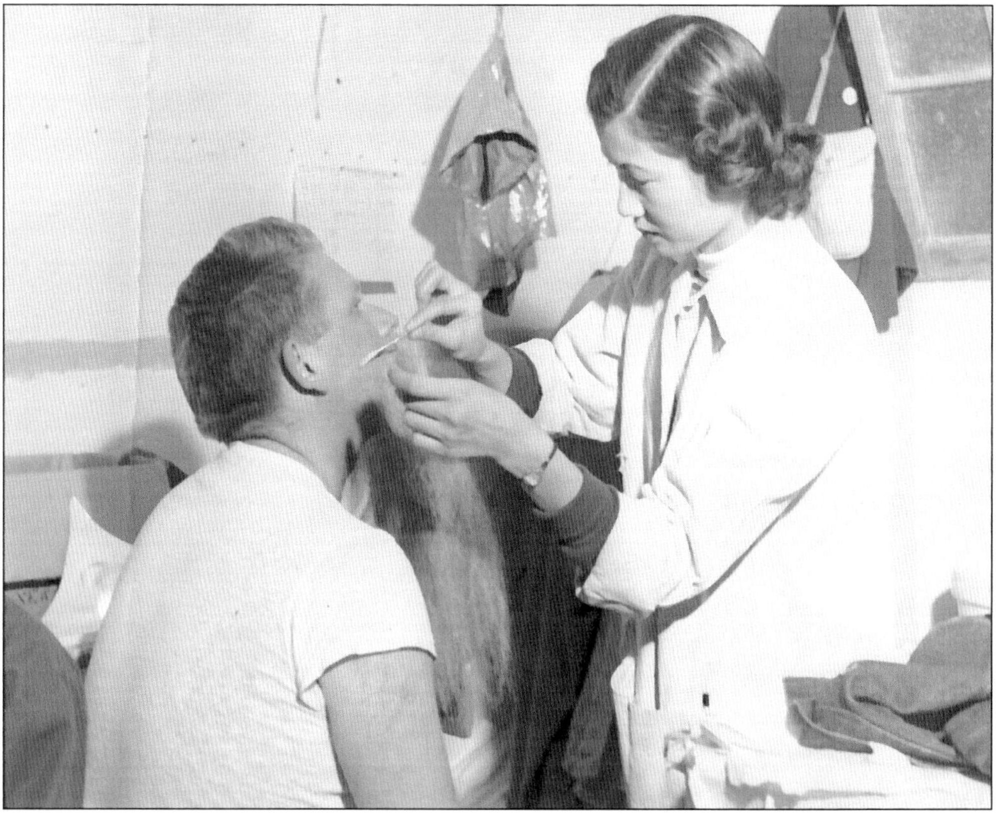

While onstage talent stole the show, backstage volunteers made the pageant possible. Above, Lily Sakemi places stage makeup on Bill Wool in an early festival, around 1953. Sakemi and her family, like so many in the Coachella Valley's Japanese community, were interred during World War II. Below, the beard transformed Wool (center), pictured here with full costume. Wool worked on the production beginning with the first pageant in 1948 until his death in 1976, both acting and working behind the scenes on the pageant board. In later performances, he often took the role of the grand vizier. (Above, photograph by Alex DePaola; below, photograph by Willis.)

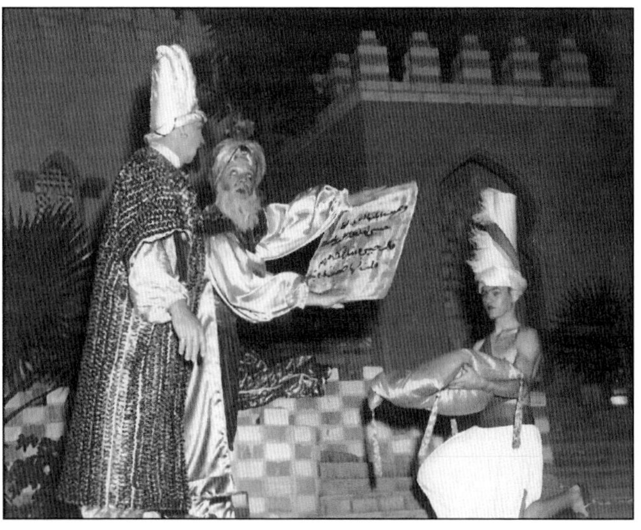

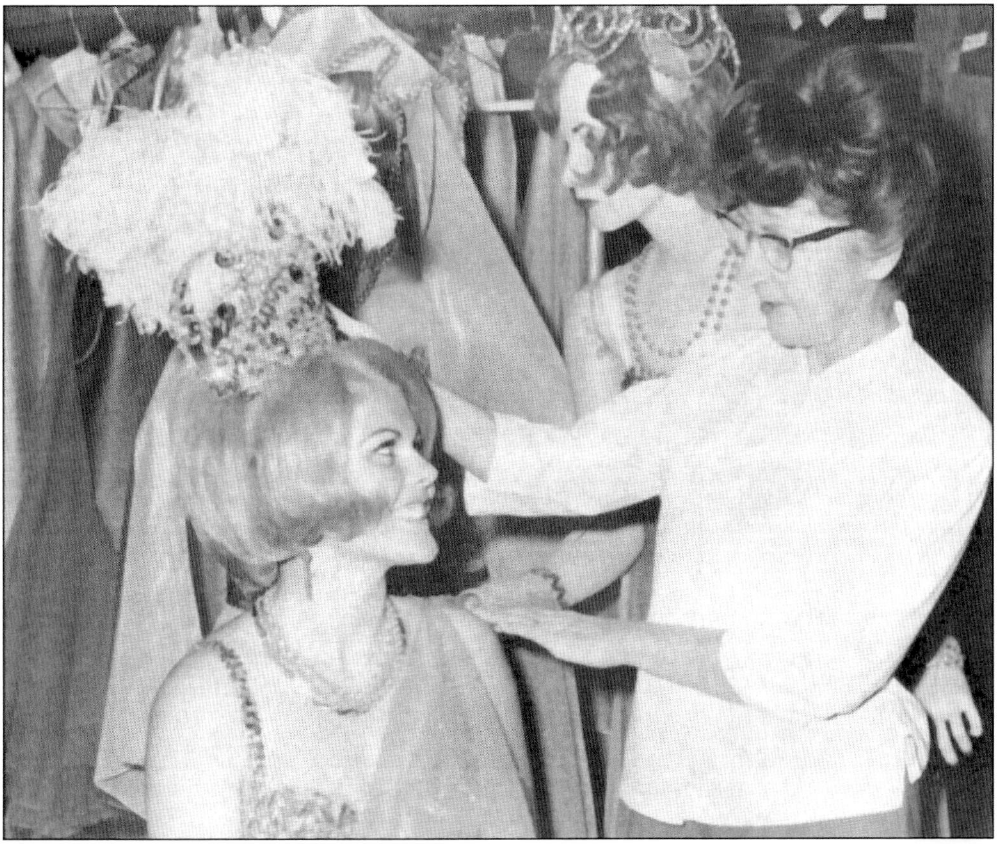

Zenda Elliott (above), pictured here with 1969 queen Donna Kennedy, headed the wardrobe department. For more than 30 years, Elliott supervised a department that designed two costumes for the queen and court, one for the evening and one for the day. Additionally, they costumed the entire pageant cast, no small feat given the changing storyline and setting every year, not to mention costume changes. Ruth Shafer served as wardrobe mistress for decades, while a dedicated group of seamstresses (right) rounded out the team, which also clothed the fair board and staff. (Both, author's collection.)

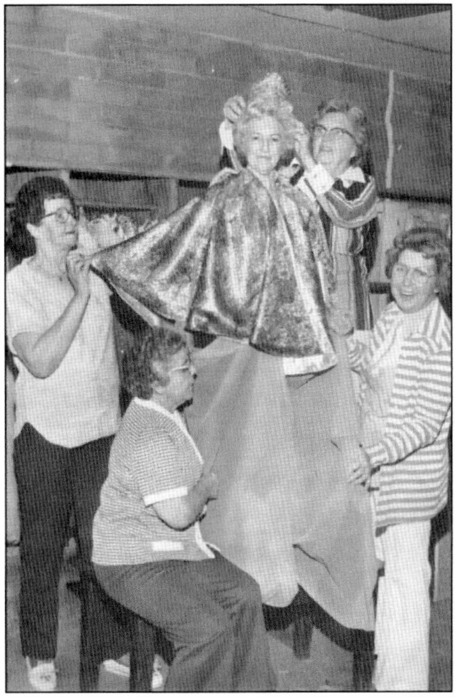

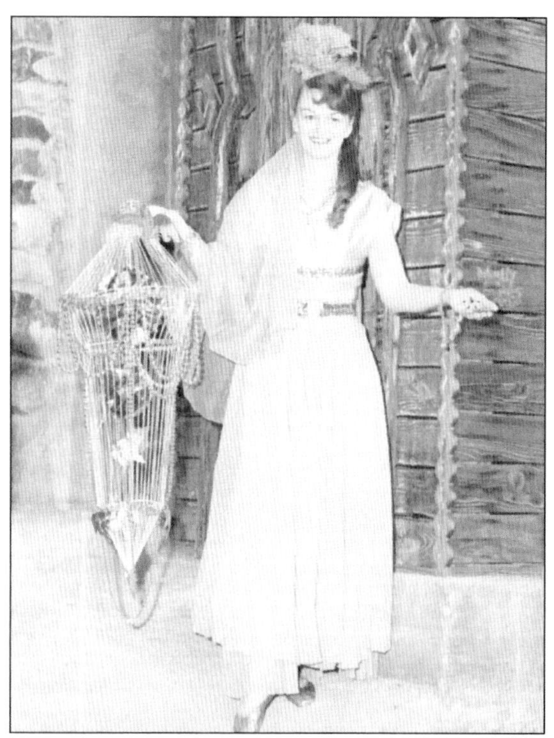

Gwen Harlow made her first appearance onstage in *Scheherazade the Lovely*, the 1952 pageant. She performed in the pageant yearly until 1998. The Harlows were active in the community; Gwen's husband, Roger Harlow, served as the mayor of Indio from 1972 to 1974 and 1981 to 1984. Above, Harlow poses for publicity photographs for the 1969 pageant, *Tale of the Enchanted Treasure*. She played the part of a street vendor. Below, Jeri Taylor (left) presents Gwen Harlow with a lifetime golden membership card in 1990 for her nearly 40 years of service. (Left, author's collection; below, courtesy of the Coachella Valley History Museum.)

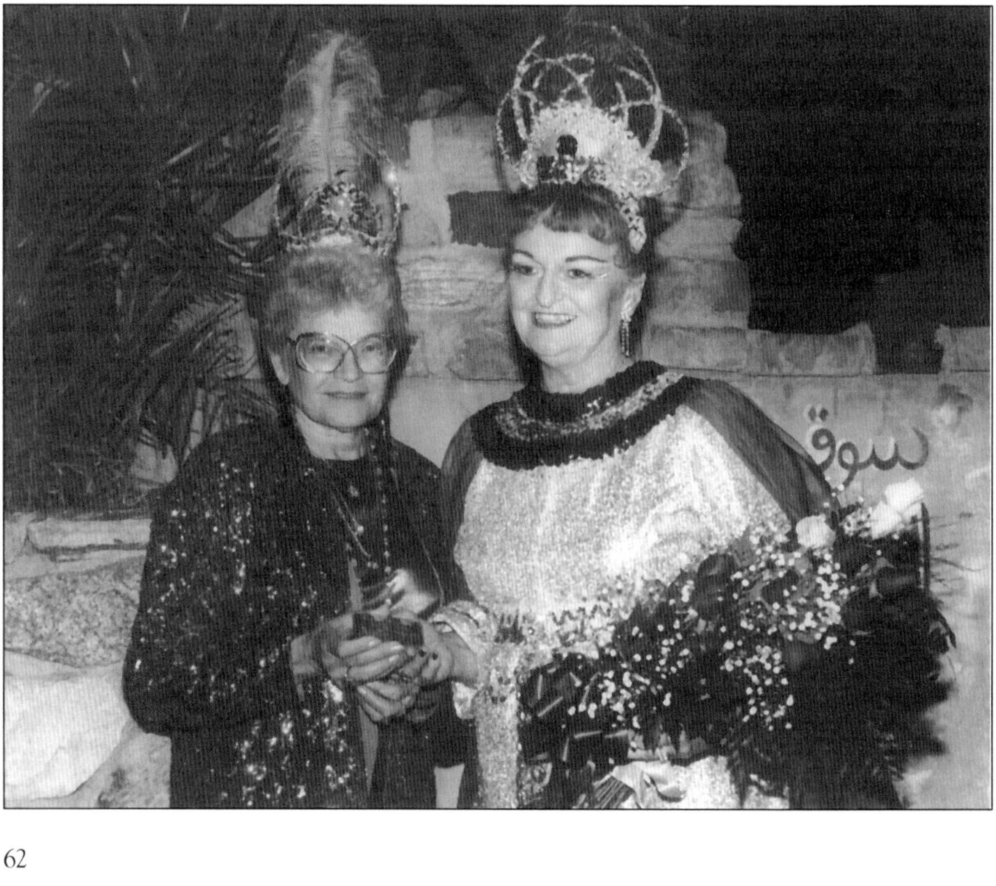

Frances Romero began dancing in the pageant in 1958 when she was just nine years old, eventually becoming assistant dance director in 1966. She leaps here in the *Tale of the Firstborn Prince*, while playing the role of Jauharah, the Teller of the Stars. (Author's collection.)

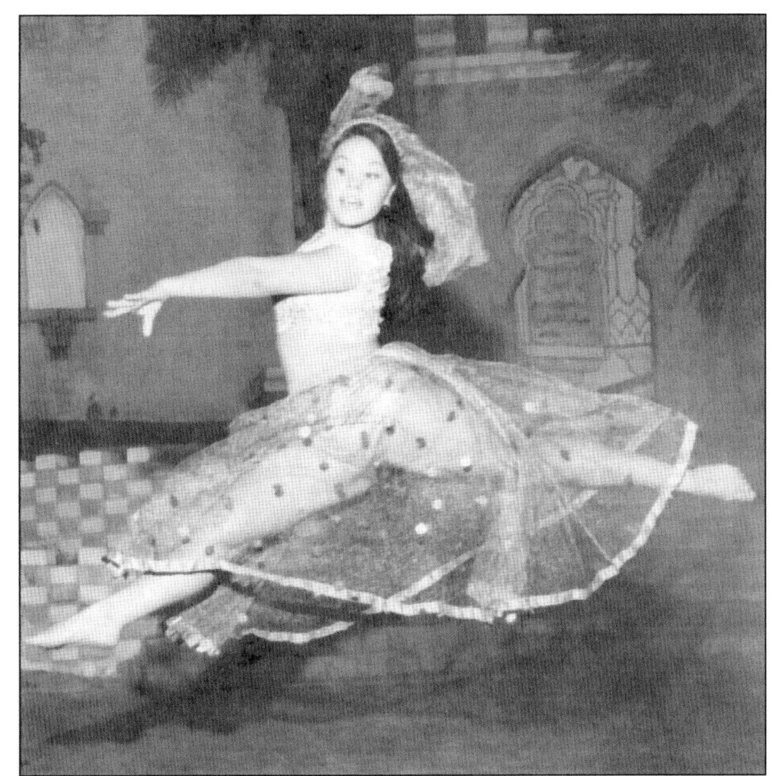

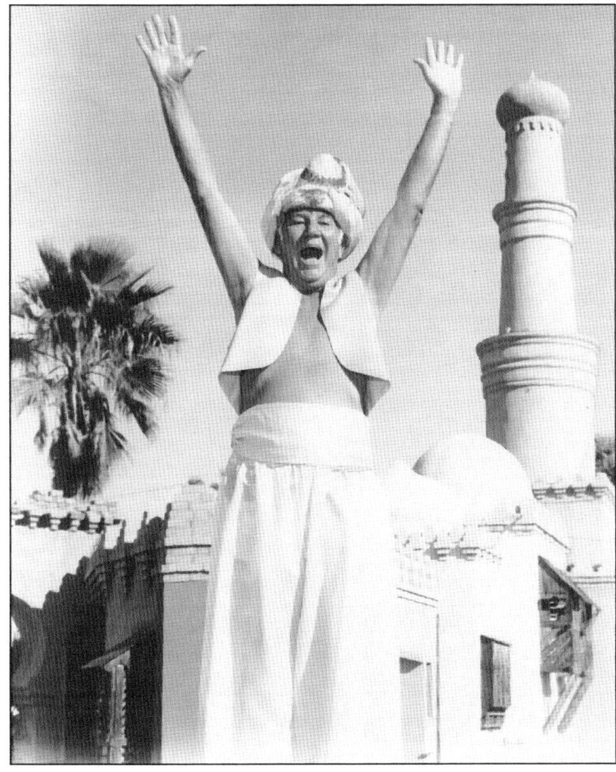

The script changed annually; not a single story was repeated until 1984. While the plotline changed, based on the countless stories of the *Arabian Nights*, audience members looked forward to certain touchstones. Nearly every play included a storyteller, caliph, and of course, a magical genie, portrayed here by Paul Nichols.

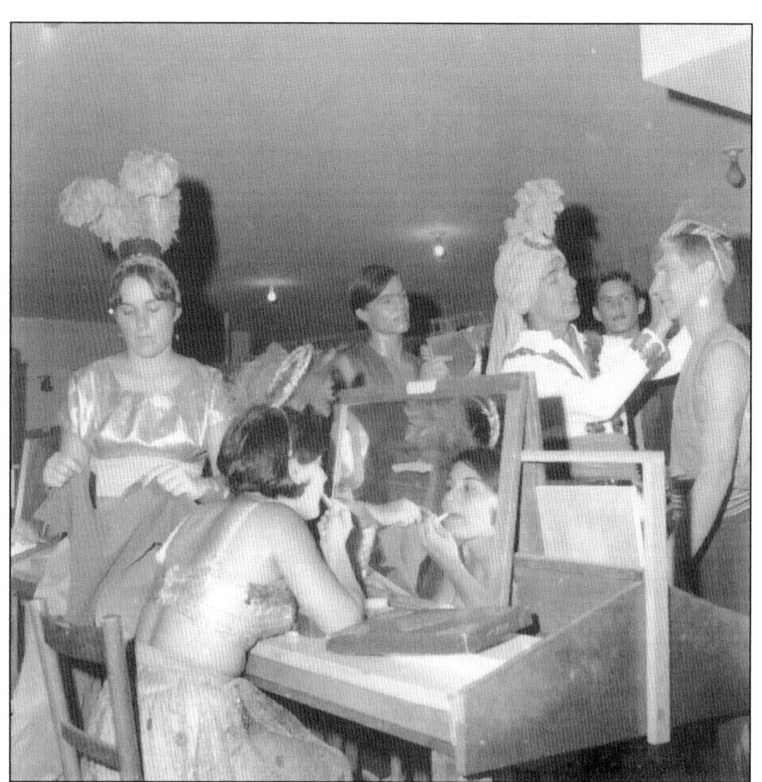

Casting normally occurred before the year's end, and rehearsals kept actors busy several nights a week. Actors, singers, and dancers sometimes had experience in Coachella Valley high school theaters, choirs, and dance teams. Performers prepare backstage in this undated behind-the-scenes photograph. (Courtesy of Palm Desert Historical Society.)

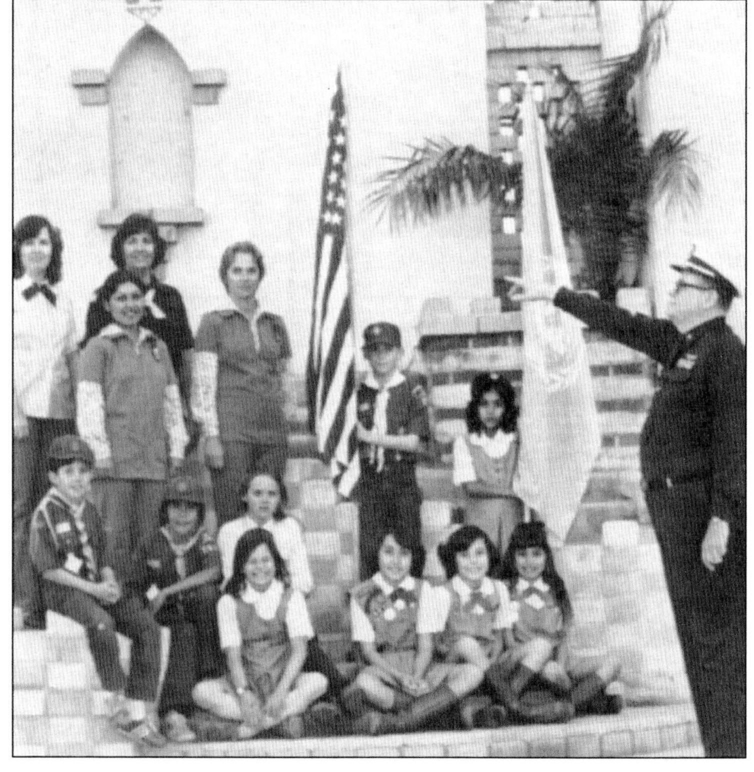

The *Pageant of the Flags* began in the 1970s. It used flags to tell the history of the United States. Originally part of the opening ceremonies, today the similar *Salute to America* takes place nightly just before the *Arabian Nights* pageant. John Power, commander in the US Coast Guard Reserve and narrator, poses with local scouts in this photograph. (Author's collection.)

Four

Date Displays
Exhibits and Attractions

Like most county fairs, the date festival featured competitive exhibits, some of which mirrored the fair's Arabian theme. Historically, displays of dates and grapefruit, the region's other prized crop, stood out in the Taj Mahal Building. There date shops displayed their best fruit and entered lavishly designed Arabian-themed exhibits. The best won ribbons and trophies, though all were rewarded with additional publicity and mail-order customers.

Local organizations such as California Women for Agriculture and the Coachella Valley Water District also educated the public about the area's agriculture and water resources via fair exhibits. Junior exhibits brought county children's projects, from miniature missions to flower arranging and landscape architecture, to a larger audience. Home cooks submitted all kinds of baked goods or jams, though the fair maintained special date categories and featured an annual date-cooking competition. The home-arts displays highlighted textiles, ceramics, handmade jewelry, and collections. Crowd favorites included the fine-art and gem and mineral shows as well as the model railroad, which now resides year-round on the fairgrounds. Other commercial exhibits also materialized at the fair, selling souvenirs and even trailers and hot tubs.

Of course, fairgoers attend for more than the exhibits. Midway attractions such as games and rides stood out in many people's memories of the festival. For others, food like date shakes and cinnamon rolls lured them in. Clowns, bands, and magic acts served as entertainment alongside the *Arabian Nights* pageant, demolition derbies, concerts, and camel races.

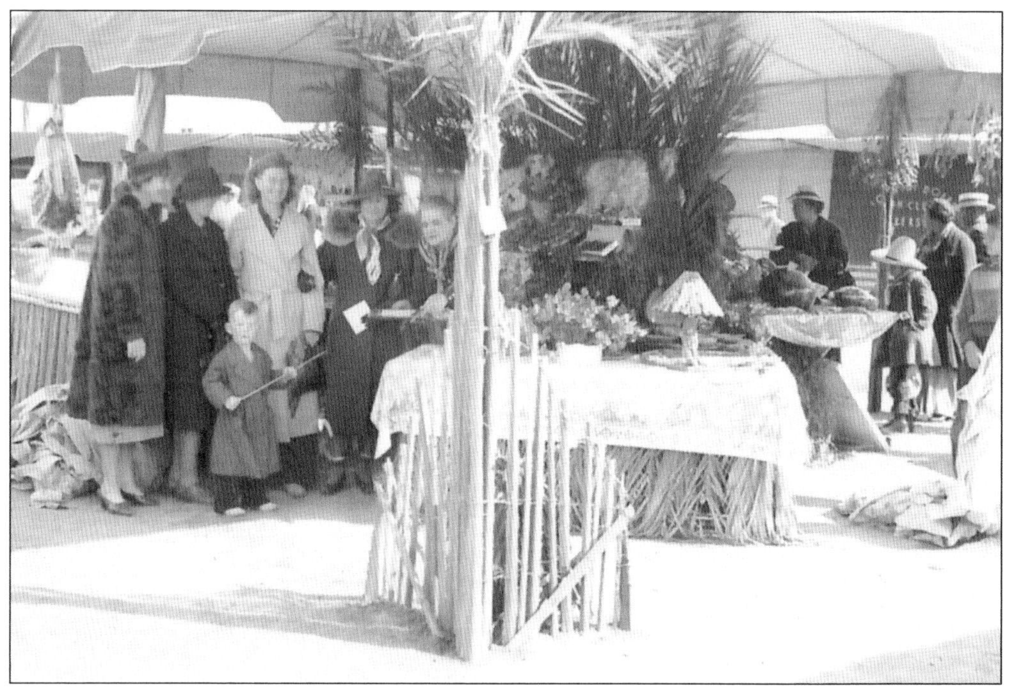

Members of the Robinson and Mitchell family pose in front of exhibits in the late 1930s. Exhibit displays from the time included homemade date and citrus goods, agricultural centerpieces, crafts, date packaging, and wildflower bouquets. Both palm fronds and date bunches were used to decorate the exhibit booth. (Photograph by Margaret Tyler.)

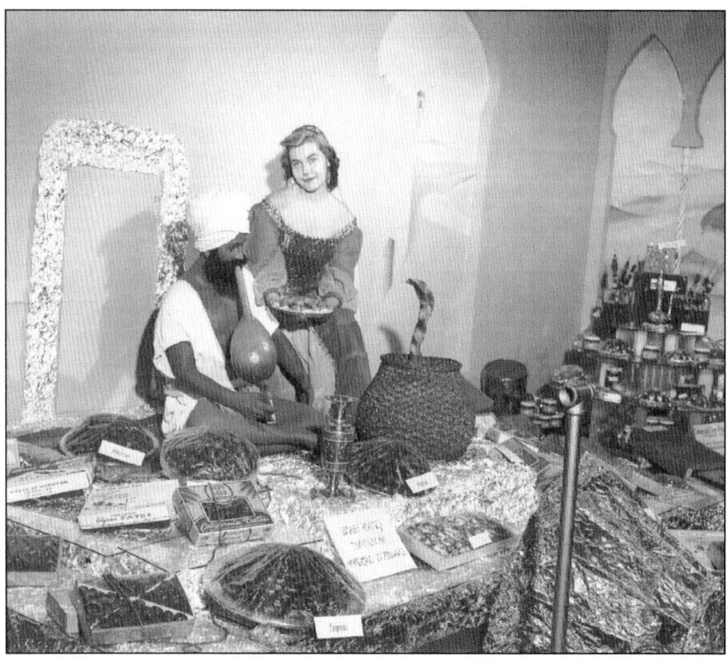

Early exhibits featured incredible details, including background murals and faux architecture, like that seen in the back right. Many also incorporated vignettes of Middle Eastern stereotypes such as this snake charmer. This exhibit included different varieties of dates such as the Medjool, Thoory, and Khadrawi. Commercial date products like date butter were displayed in the back right. (Photograph by Roy Gillman.)

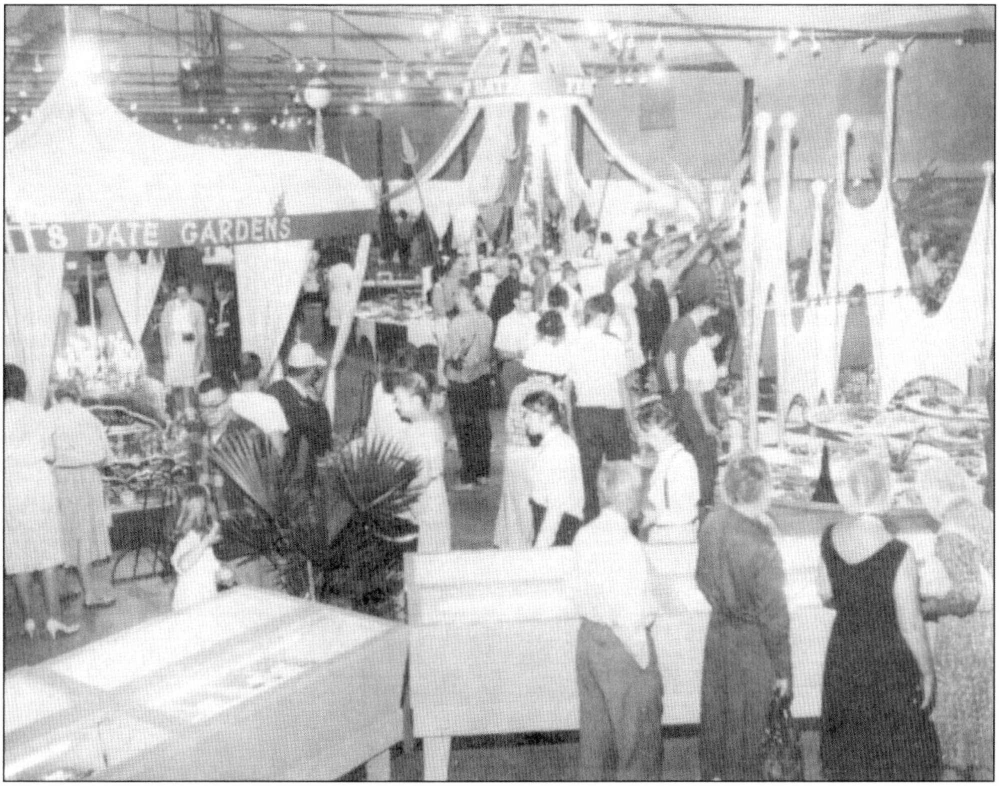

Exhibits proved popular with the public. Sniff's Date Garden (left), the California Date Growers Association (center), and Valley Date (right) created large exhibits to display different varieties and products offered. Entrants won cash prizes for best exhibits and best dates, which were plated and displayed. (Author's collection.)

The Laflin Date Garden exhibits won prizes annually, not only for their use of the Arabian theme but also for their educational nature. The 1962 display recalled the date's journey from the Middle East via Spanish missionaries and the US Department of Agriculture. The queen cradles a male spathe, which holds the pollen needed for female palms to produce dates. Typically, growers plant one male for every 49 females. (Photograph by Roy Gillman.)

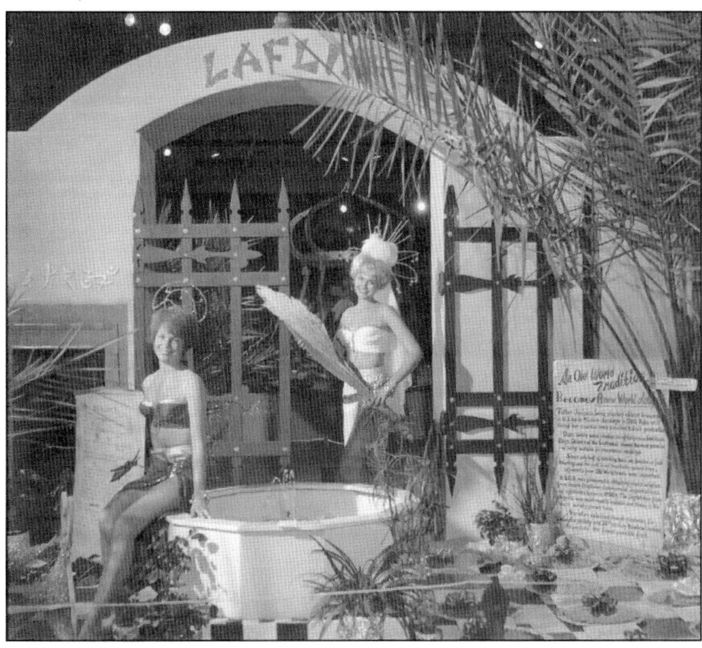

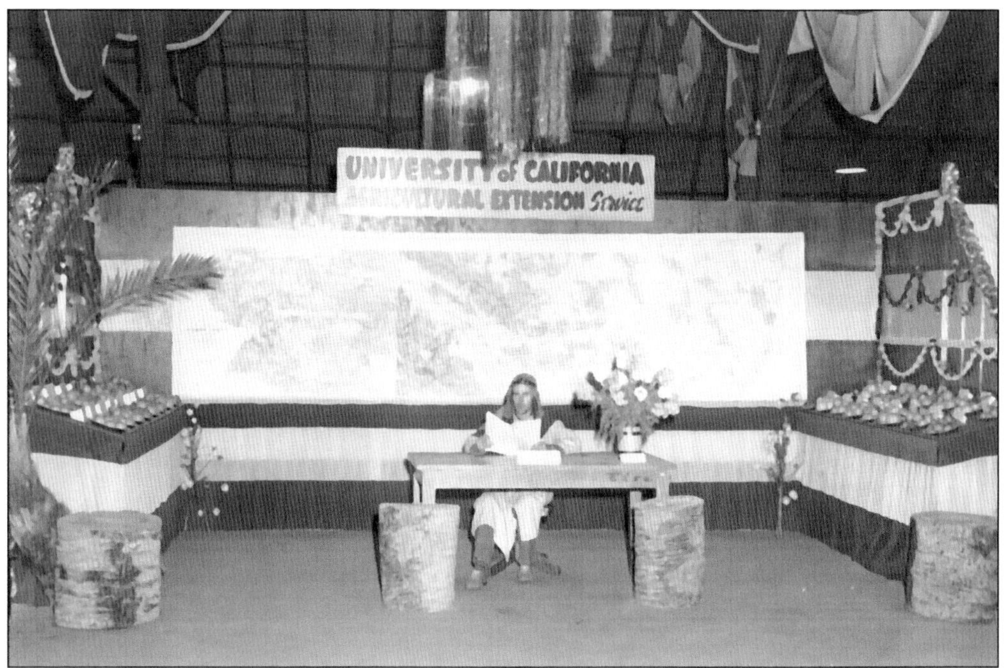

After its creation in 1954, the University of California, Riverside, led the way in date and citrus scientific research, alongside the US Department of Agriculture. The University of California's Extension Service (as well as the Riverside Experiment Station) predated the university, providing scientific support to farmers in the region. The top photograph features a costumed employee ready to educate the fair visitor around 1949. The map of Riverside County behind him pinpointed where various crops (dates, cherries, and potatoes) grew best. The exhibit also featured different citrus varieties seen on the wings. Below, another extension-service exhibit focused on the pests that harmed the date crop. On display were fungi and insects trapped inside the beakers. The exhibit also included the same map and samples of regional soil and wheat. (Both, courtesy of Heidi Hutchingson.)

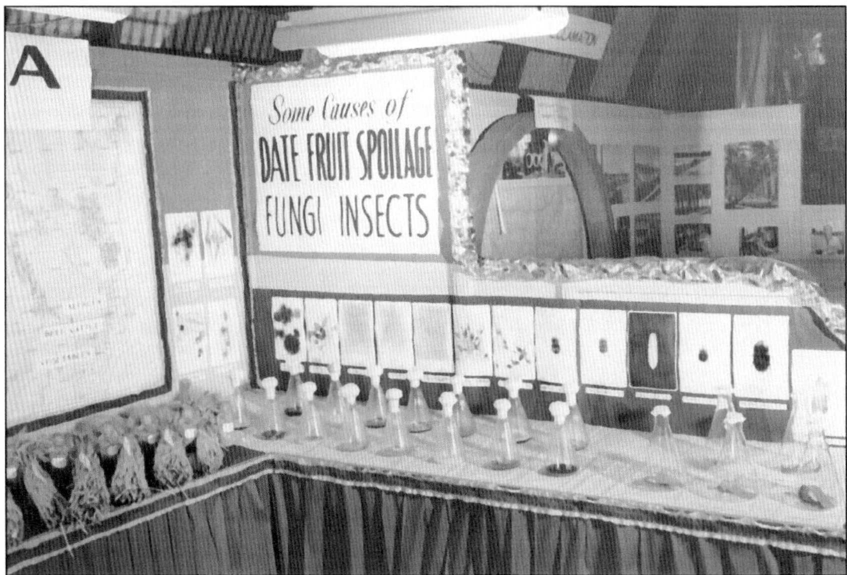

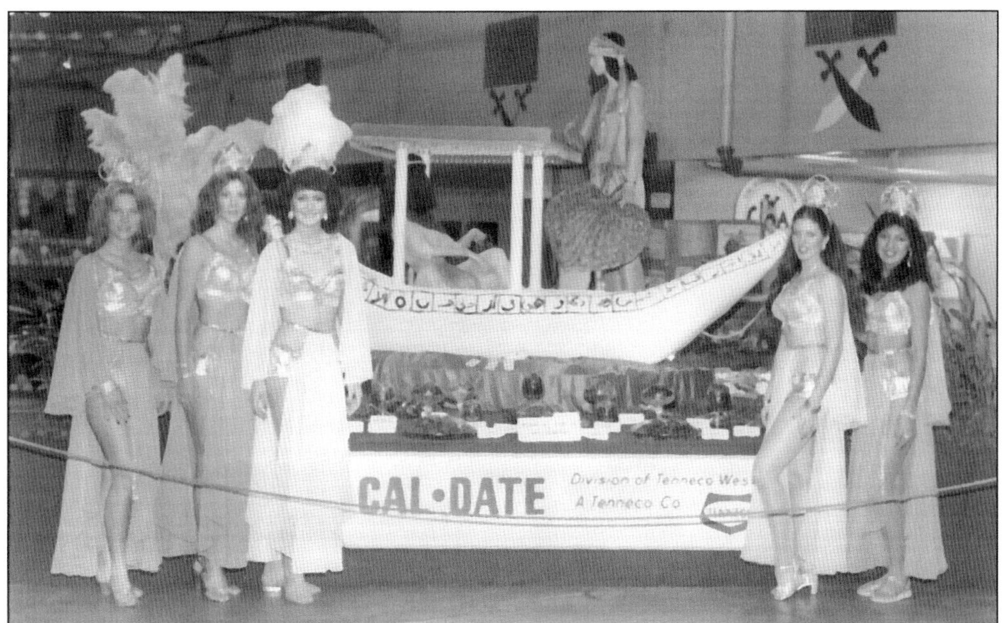

For much of the fair's history, local date shops vied for space in the date exhibits building, hoping to increase date sales through this outreach. Shops such as Sniff's, Covalda, Jensens, Shields, Valley Date, and the Laflin Date Shop drew acclaim for their exhibits. As the 20th century wore on, many of these mom-and-pop date shops folded, and agribusiness took over. Above, Tennco's Cal-Date exhibit houses the entries for the date competition. Note the decoration in the Taj Mahal; streamers and fabric adorn the ceiling while crossed scimitars decorate the walls. Below, the 1985 Covalda display incorporates a pyramid with an inlay design made from dates. Lee Anderson, the owner of Covalda, pioneered the use of brown paper bags to protect ripening dates from the rain. (Above, courtesy of the Coachella Valley History Museum; below, courtesy of Catherine Anderson Kerby.)

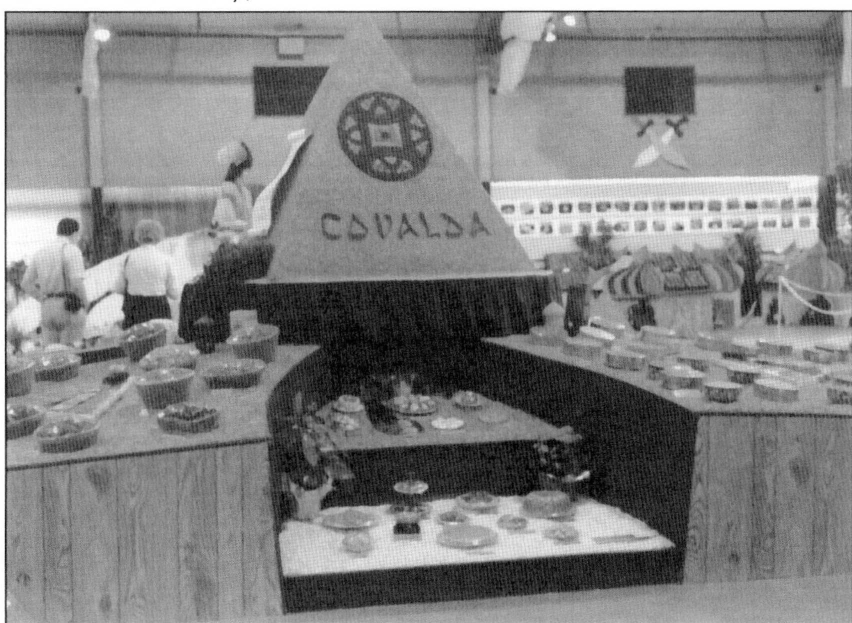

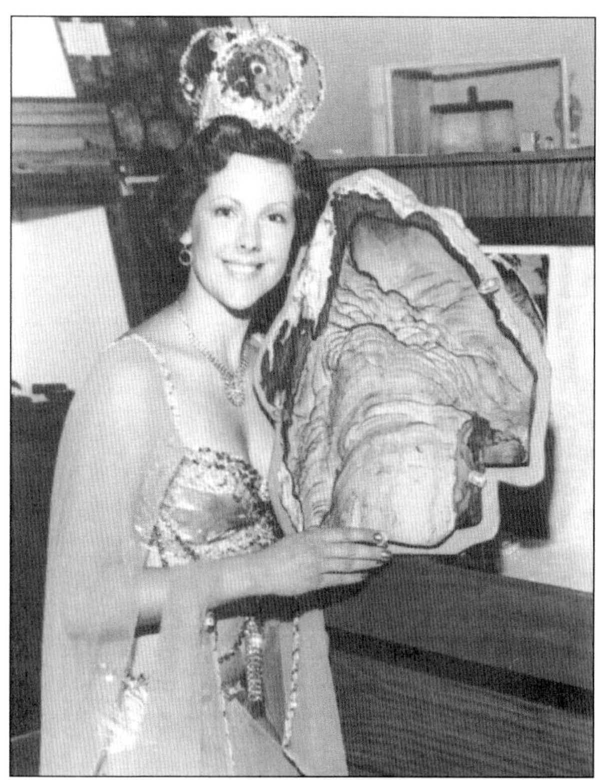

The gem and mineral show drew attention from rock hounds around Southern California. Volunteers from regional gem and mineral societies polished rocks and educated the public while vendors sold gemstones, jewelry, and beads. Displays featured precious stones and fossils, an audience favorite. Other exhibits included Native American artifacts, mining equipment, and even a table set with rocks that looked like food. The advanced-division exhibits were open to any collector in the world, and the show was considered one of the best in the United States. (Both, author's collection.)

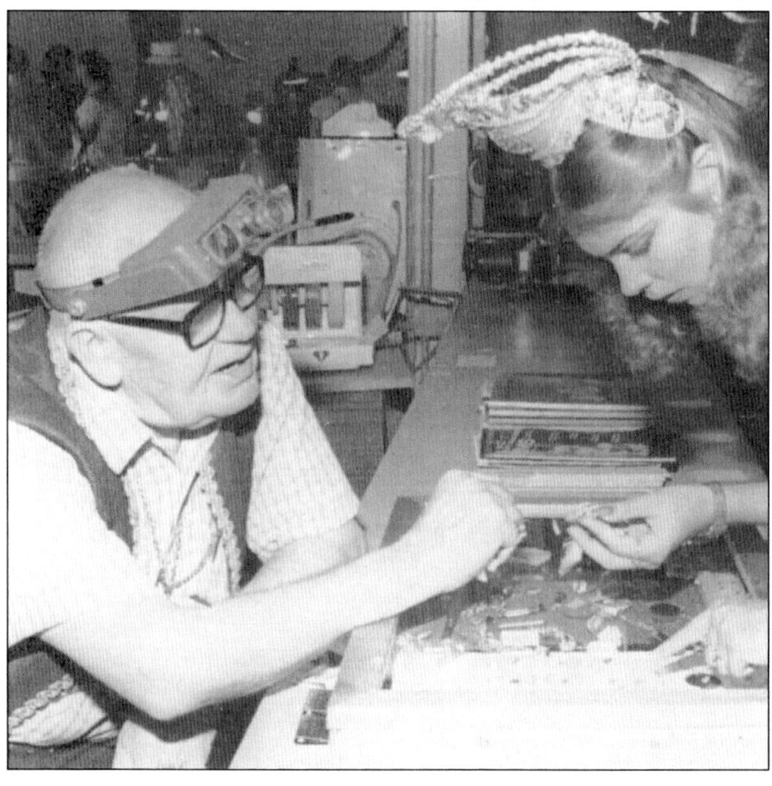

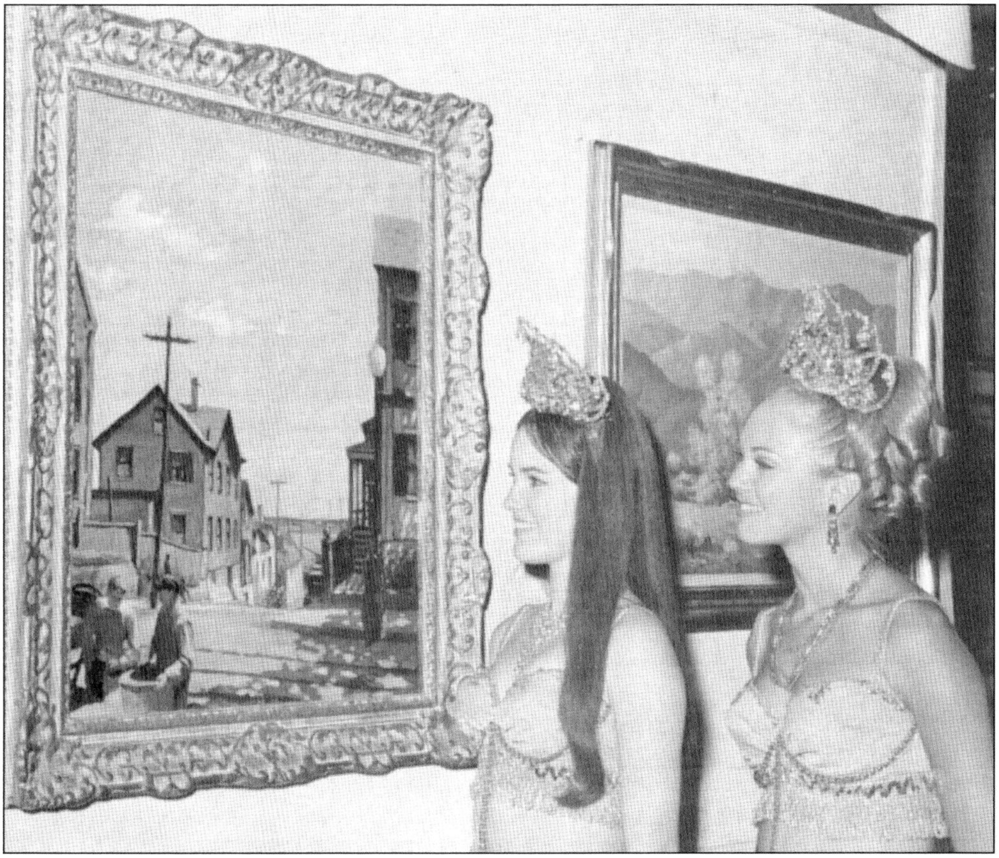

The fine-art exhibits allowed artists, amateur or professional, to display works to a receptive audience. Official judges distributed ribbons, but fairgoers also voted for their favorites, with a cash prize going to the most popular piece. Above, members of the court explore the fine-art show in 1969. Below, John Hilton displays his work, about 1955. Hilton operated a gem store across from the famed Valerie Jean's Date Shop near Thermal, California. One of his works, *Twentynine Palms Oasis*, hung in the Oval Office during Eisenhower's administration. Hilton recalled meeting Eisenhower, a fellow painter, at one of his shows in Palm Springs when Eisenhower was still a general. They kept in touch; Eisenhower invited Hilton to his inauguration, and Hilton presented him with the painting in 1957. (Above, author's collection; below, courtesy of Linda Beal and the History Hunters.)

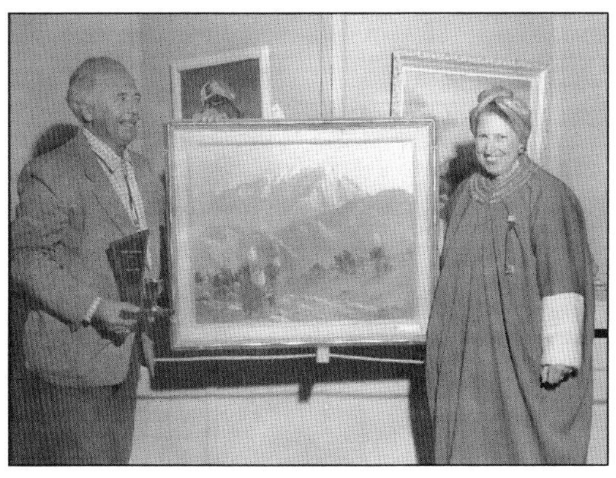

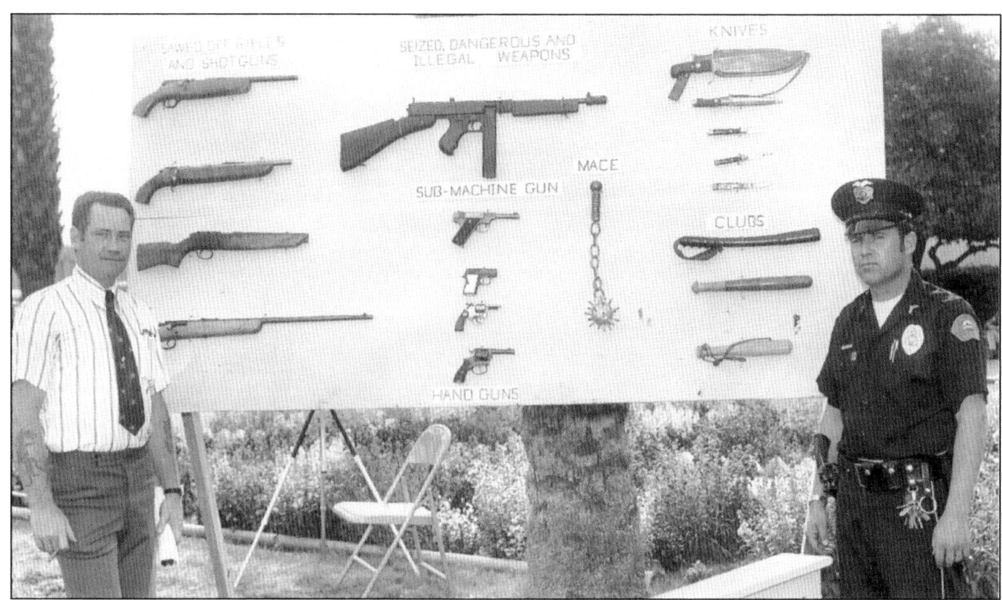

Local cities, departments, service clubs, and nonprofits utilized the date festival for public outreach and educational displays. Above, in 1972, the Coachella Police Department displays seized weapons, including sawed-off shotguns, a submachine gun, and clubs. Chief Lester O'Neil stands on the right, opposite Det. Bill McMullin. (Photograph by Walt Frisbie, courtesy of Lester O'Neil.)

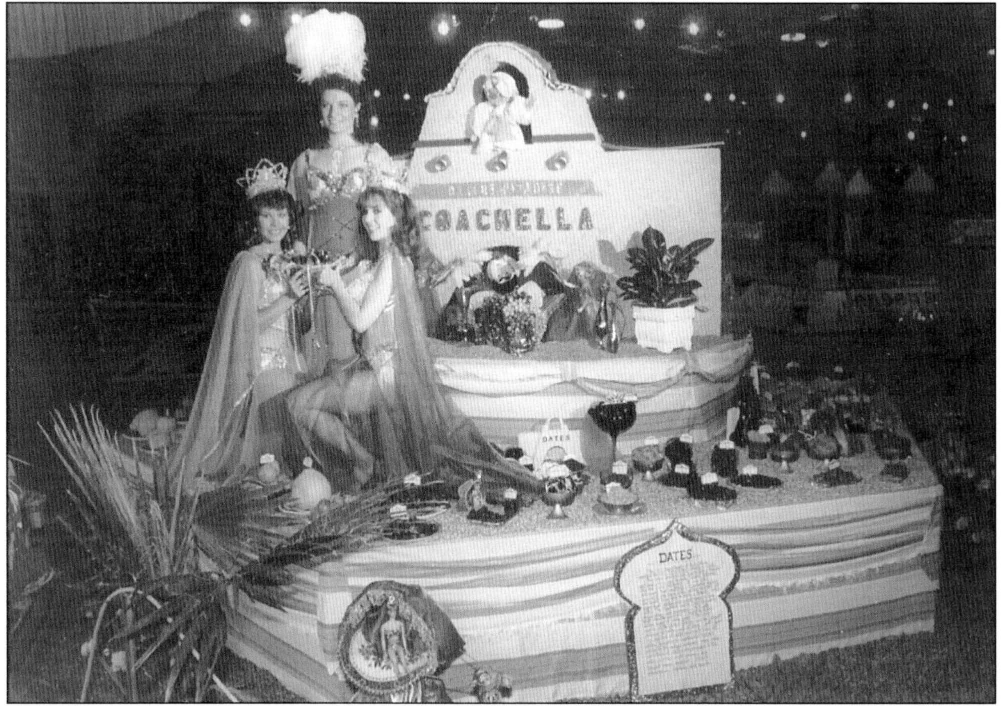

Various cities in the Coachella Valley and Riverside County contributed exhibits to the fair, hoping to draw business, tourism, and residents to their town. Here, the Coachella exhibit features Barbies dressed as harem girls riding camels (bottom center) as well as grapefruit, grapes, and dates, the city's main cash crops.

The Mexican Coordinating Council hosted educational exhibits that showcased Mexico in hopes of fostering goodwill between the two nations. Mexico long played a role in the date festival, sending officials to the first Festival of Dates in 1921. Mexican officials continued to attend Salute to Mexico events at the fair. Above, the floating gardens of Xochimilco, located just outside Mexico City, were recreated for this display. Historically, the *trajineras*, gondola-like boats used in the canals, were adorned with flowers. Here, a young man mixes in a bit of the Arabian theme with his mustachioed sultan costume. Below, Gilda Guitron Sandsness (left) and her sister Elisa pose with a Mexican dignitary in the goodwill booth. The exhibit featured artisanal goods, including pottery and leatherwork crafted in Mexico. (Above, author's collection; below, courtesy of the Gilda Guitron family.)

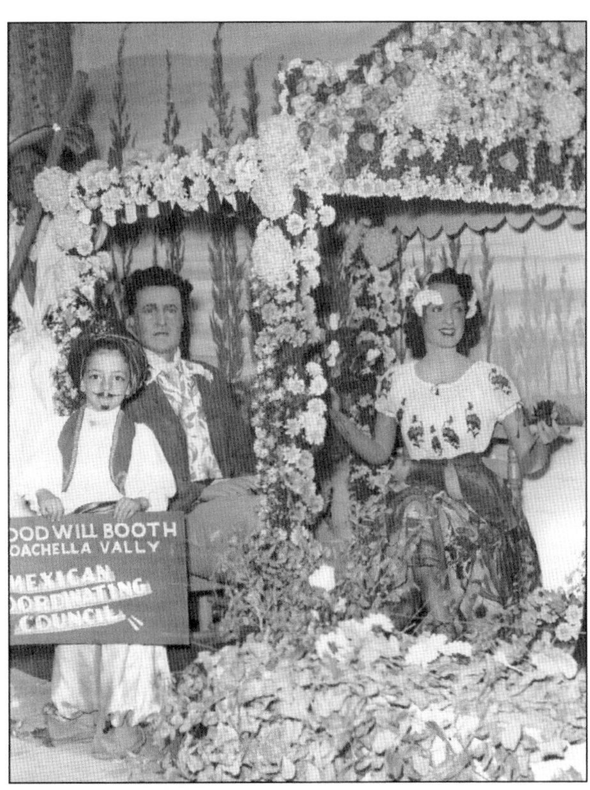

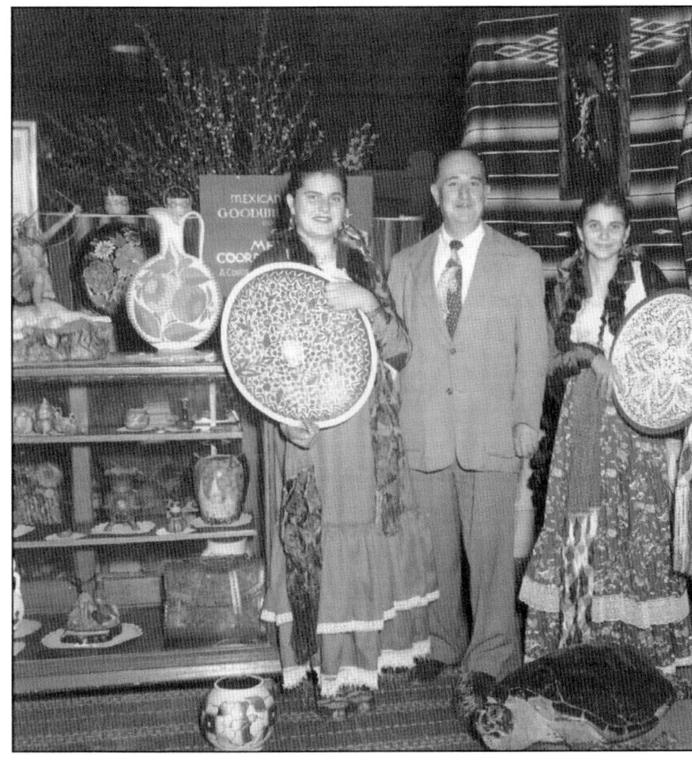

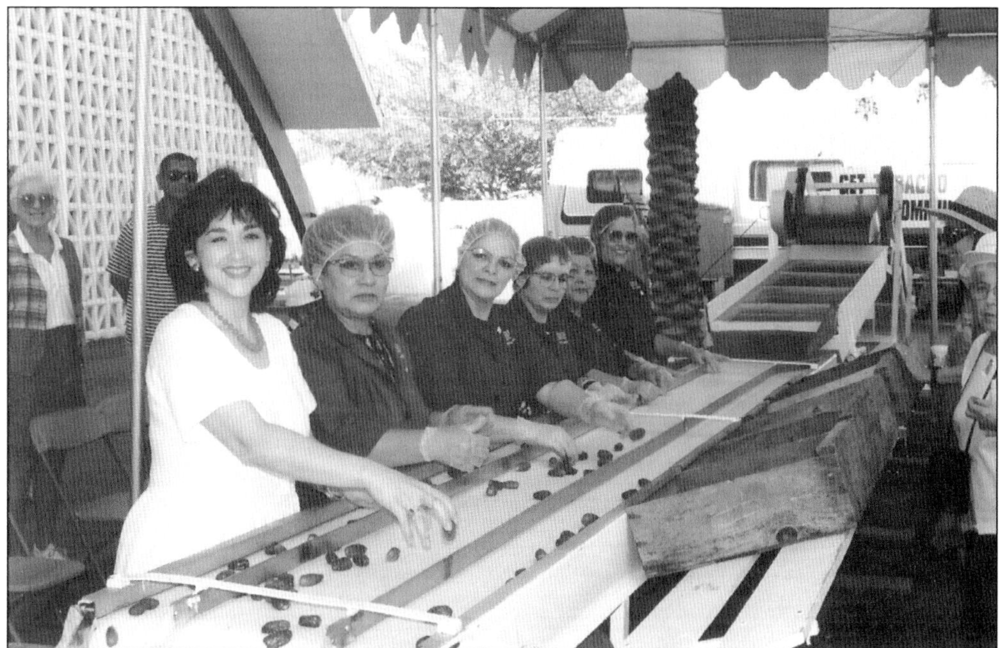

While the fair's Arabian theme referenced the crop's homeland, actual date displays drew delight from visitors. Here, the team from Hadley Date Gardens demonstrates date grading and processing in front of the Taj Mahal Building around 1997. (Courtesy of Hadley Date Gardens.)

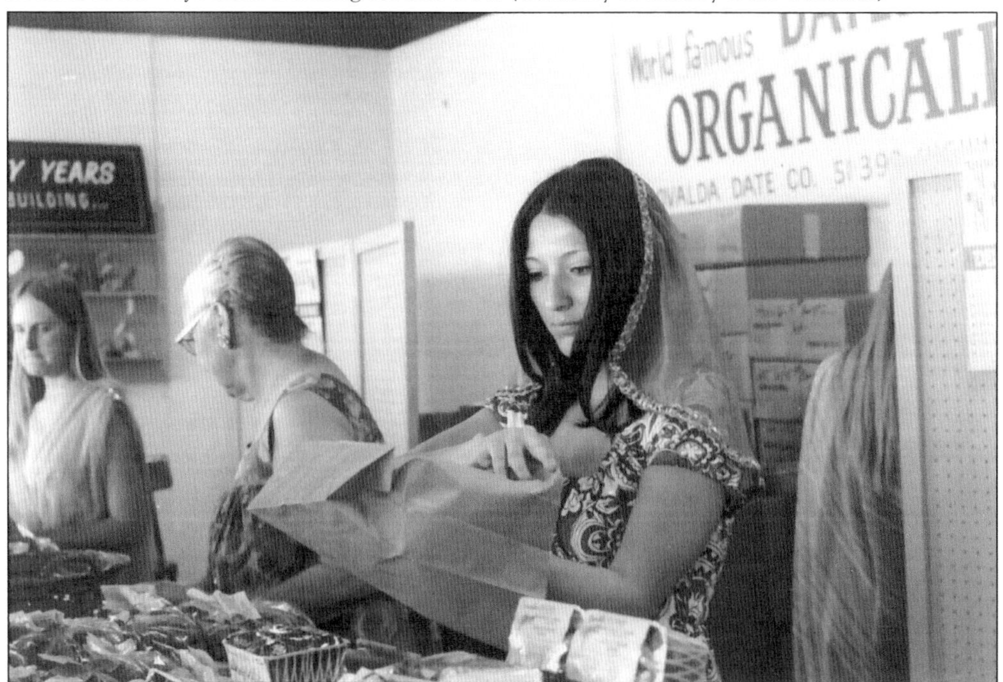

Fairgoers arrived from all over Southern California. Many stopped to buy dates along the drive, but others waited to make their purchases at the fairgrounds. Laflin Date Garden (later Oasis Date Gardens), Sniff's Date Gardens, and Covalda (pictured here in 1971) sold their dates and date products at the event. (Courtesy of the Lee Anderson family.)

The midway fair rides have long been a favorite. Festival organizers hired touring companies that operated the rides, though for some time the ticket sales supported local clubs and nonprofits. Above, princesses ride one of the roller coasters set up on the fairgrounds. The Ferris wheel, seen in the background, was invented by George Washington Gale Ferris Jr. for the 1893 World's Columbian Exposition in Chicago. That exposition also debuted the term midway. Below, kiddie rides enticed many families to return to the fair annually. This diesel-powered train delights a young buckaroo at the fairgrounds. (Right, author's collection; below, courtesy of David Eppen.)

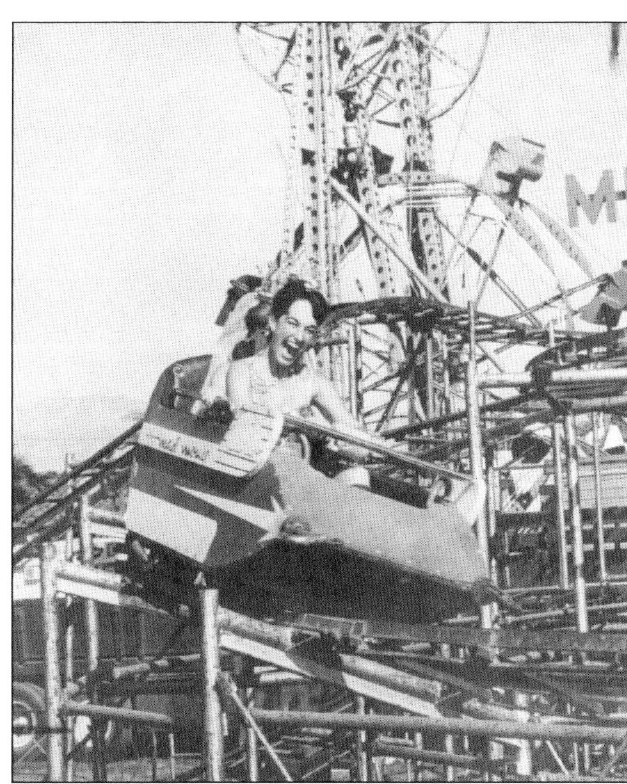

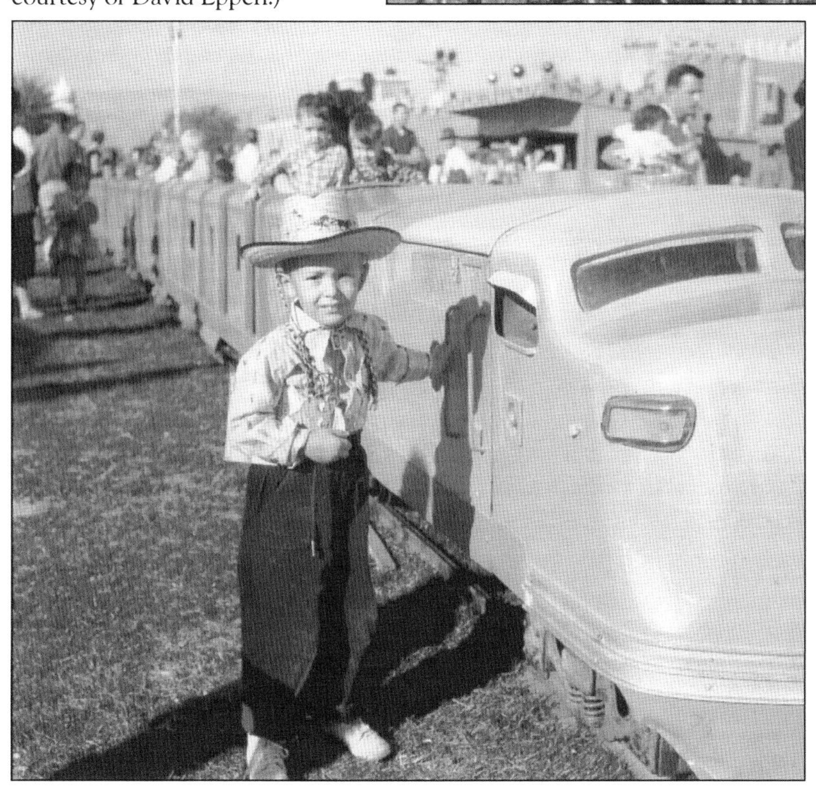

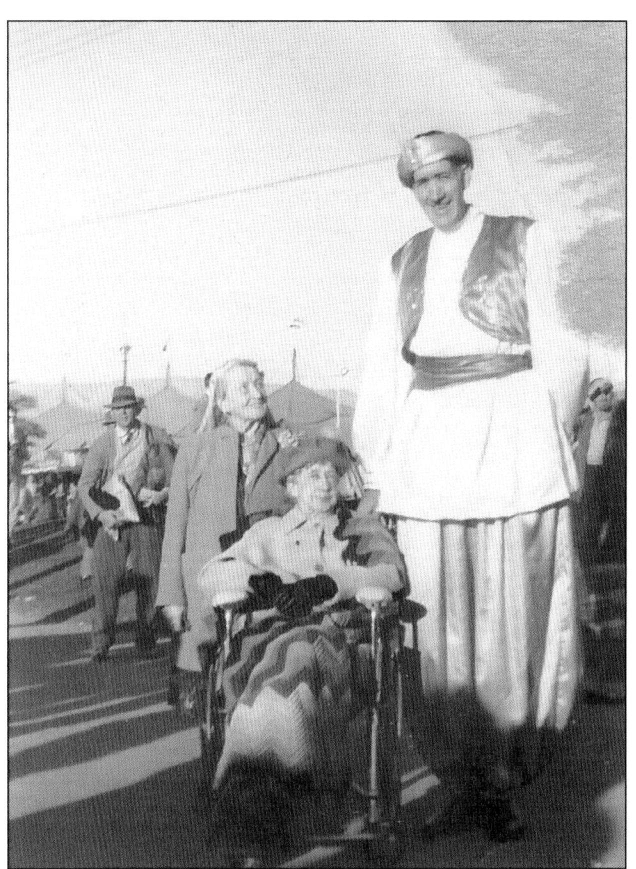

Sideshows drew attention at the festival. In 1953, sideshow performer Martin Lawrence of Chicago, Illinois, stood over fairgoers at seven feet, seven inches. Some concessionaires and performers like Lawrence donned Arabian-inspired costumes. (Photograph by Ann Clifford, courtesy of Linda Beal and the History Hunters.)

Games like the dime pitch, fish bowl, duck pond, milk bottle, balloon and dart, and water-gun races delight visitors of all ages even today. The 1985 court embraces the large puppy prizes available at this ring-toss game.

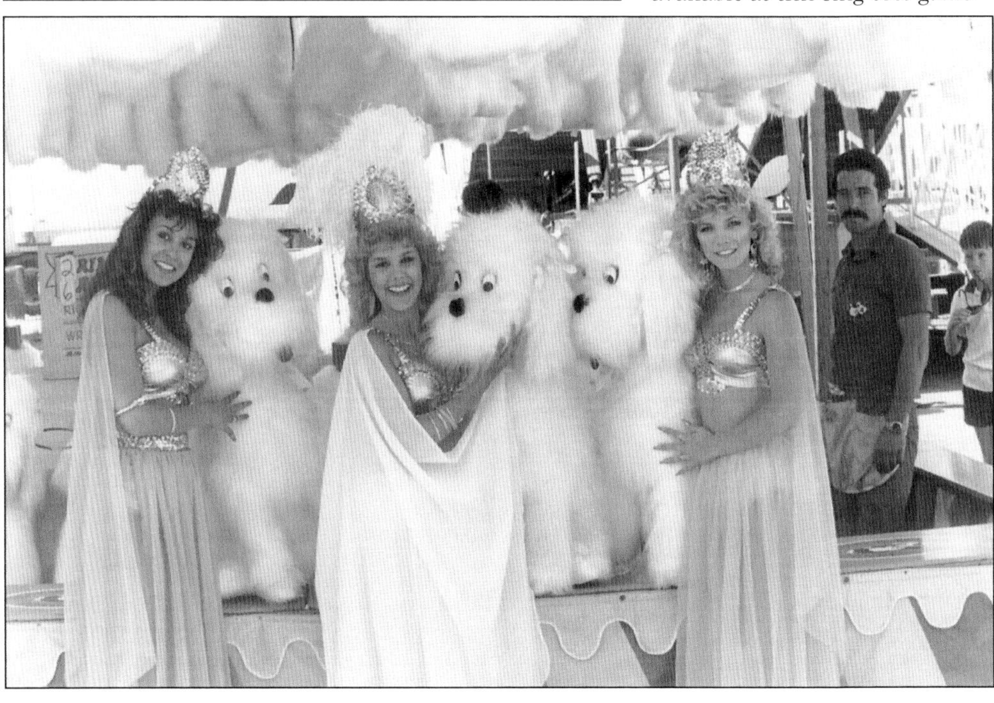

Fairgoers looked forward to the food every year. Hot Dog on a Stick began in Santa Monica, California, in 1947, expanding by offering its products at county fairs in Southern California, including the date festival. Date items were harder to find, though date shakes were always on hand. In 2013, Pink's Hot Dogs offered an Indio Date Dog at the fair, complete with bacon and chopped dates.

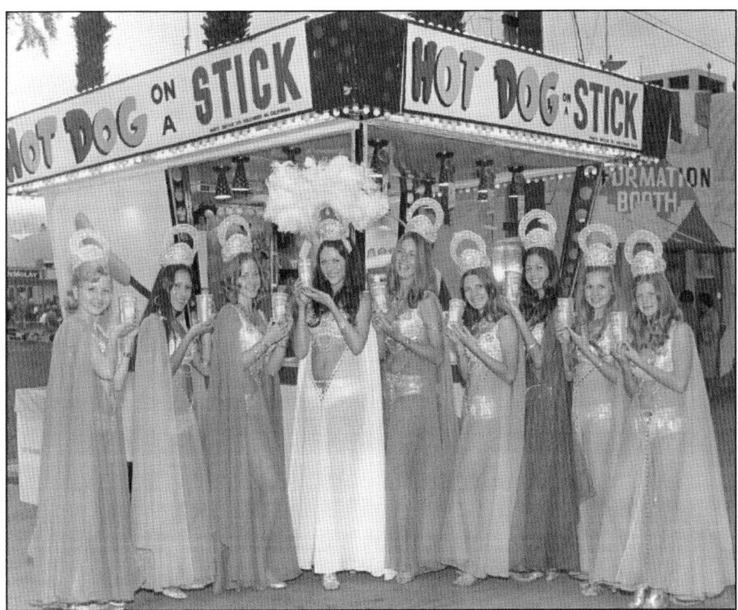

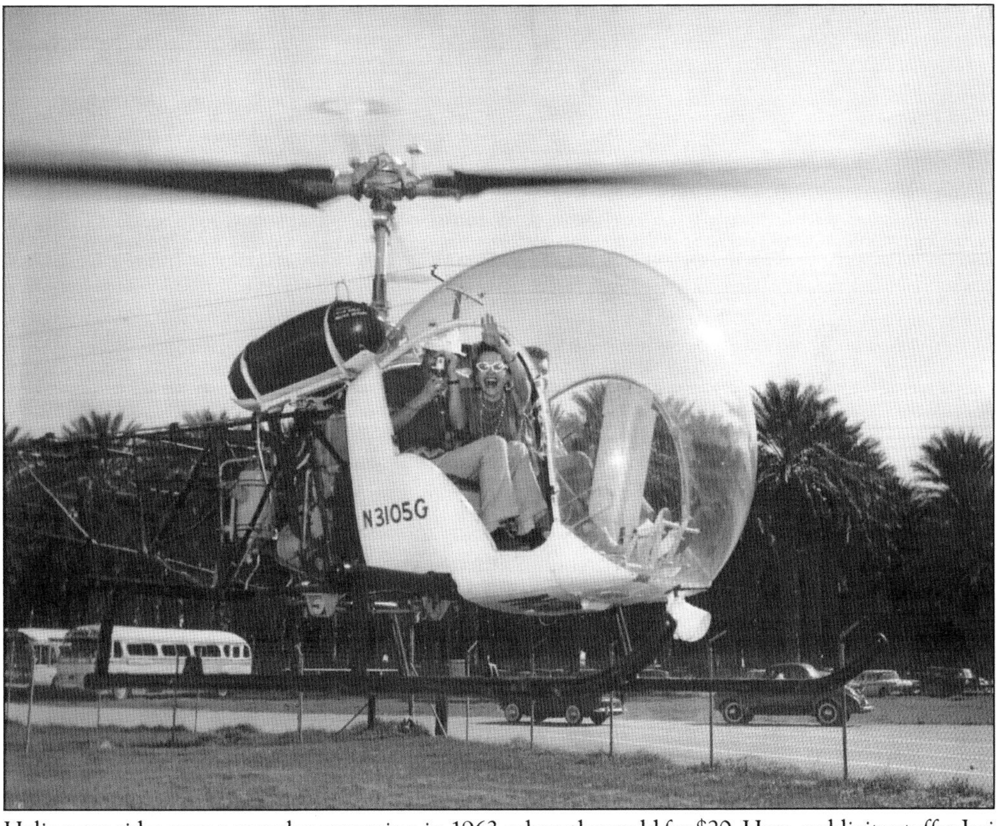

Helicopter rides were a popular attraction in 1963, when they sold for $20. Here, publicity staffer Jeri Taylor waves during her sightseeing tour. On February 24 of that year, the helicopter crashed just outside the fairgrounds. Two Southern California women were killed within sight of their families. The pilot, thrown from the helicopter, was hospitalized in critical condition but survived.

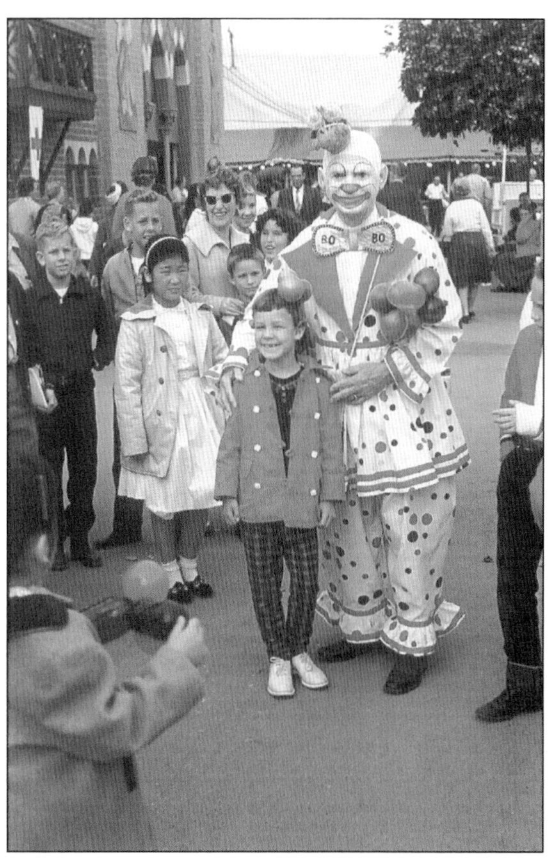

Clowns delighted children of all ages on the midway. Here, BoBo the clown entertains a gaggle of children in front of the Palace of Commerce. (Courtesy of David Eppen.)

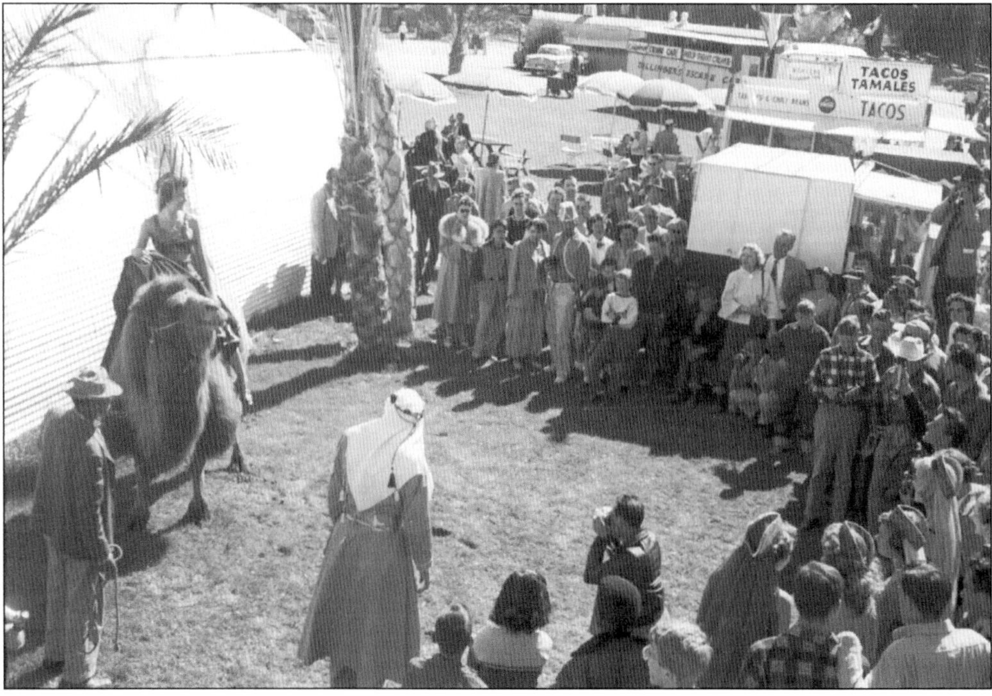

Photography events were popular at the fair. The queen and court took press photographs in the days leading up to the fair, but they were also available daily for amateur photographers. In this image, a princess poses with a camel, while several members of the audience take photographs. The background depicts additional attractions: John Dillinger's escape car and a taco and tamale stand. (Photograph by Glenn Stocking, courtesy of the Schleis family.)

Five

Camel Craze
The Animals of the Date Festival

While the human visitors at the fair drew attention with their elaborate Arabian-inspired costumes, the exotic nature of animal guests also delighted the public. The first public camel races in the United States were held at the fairgrounds in 1947. Fair manager Robert Fullenwider thought a camel might bring the fair additional publicity, so he contacted the San Diego Zoo. The zoo referred him to the World Jungle Compound in Thousand Oaks, California, where staff suggested that renting four camels and racing them might garner even more publicity. While elephants are not native to the Middle East, they also drew visitors' attention at the fair, a hint of the greater Orient and a curiosity unto themselves.

Aside from the circus, the fair was one of the few opportunities to see both of these massive creatures. They appeared in the parade and onstage in the musical pageant and accompanied the princesses to many photo shoots, with their images reproduced nationwide. As the fair grew, so too did the appearance of other animals as attractions: ostrich races, shark tanks, sea lion shows, dancing monkeys, petting zoos, pig racing, and, of course, the goldfish carnival games.

Domesticated animals also temporarily resided in the fairgrounds. Horseshows drew large crowds. Indeed, horse racing made the date festival possible; in the 1930s, new laws legalizing pari-mutuel wagering provided funding (via licensing fees) for fairs around the state. Fair attendees could place their bets at the Shalimar Building after it opened in 1975. There is more to the fair's menagerie than camels, elephants, and horses. As an agricultural fair, the local community has long shown their prized livestock, particularly through youth programs such as 4-H and Future Farmers of America. These programs allow students to raise farm animals—from chickens to cows—which are then judged and sold at auction, encouraging interest in agriculture among the fair's youngest visitors.

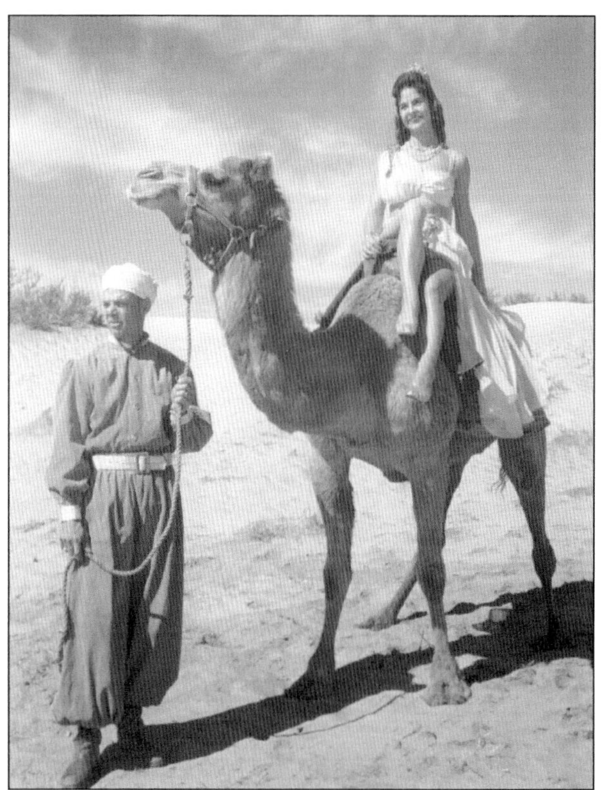

Camels made for an ideal prop, instantly signaling a Middle Eastern theme to the world. Newspapers all over Southern California reproduced images like this one. Queen contestants often participated in photo shoots a few days prior to the fair, posing both at the fairgrounds and in the sand dunes of nearby deserts.

The date festival celebrates an Arabian theme, though very little is authentically Middle Eastern. Sometimes the fair even got the camels wrong, mostly due to limited supply. Early festivals employed two-humped Bactrian camels, natives of the Gobi Desert and widely used as pack animals in Asia. They were used in camel races until 1952, when they were replaced by the faster dromedary camel.

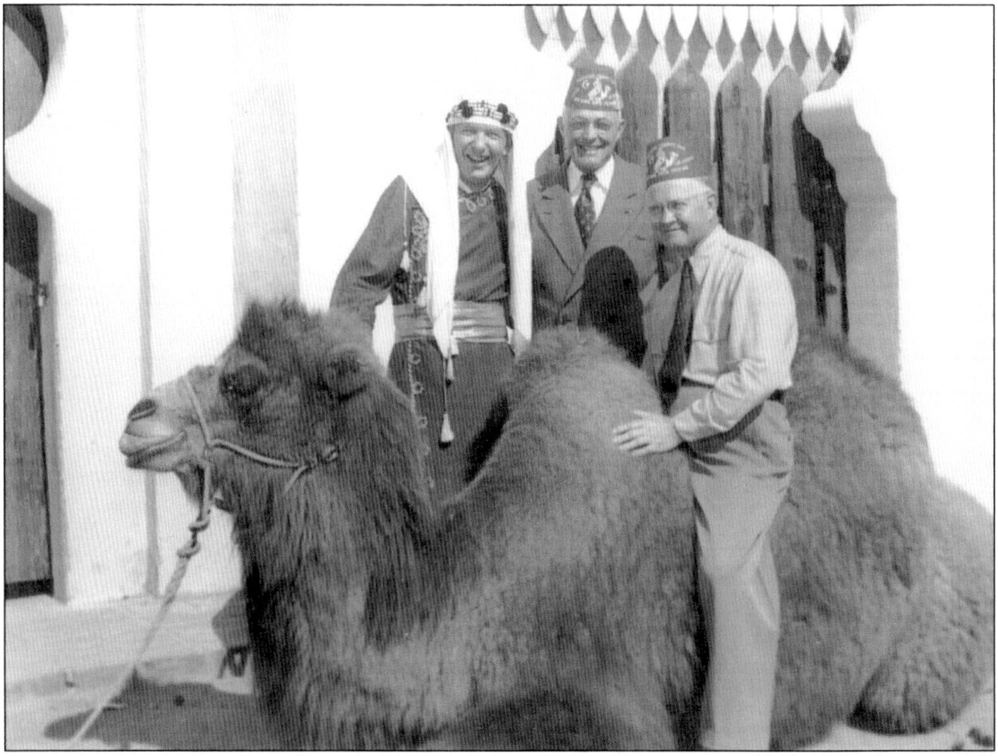

Eventually, the fair switched to the more regionally appropriate one-humped dromedary camel. Manager Robert Fullenwider wanted to purchase camels in the early years of the fair and sent word to a US Department of Agriculture date scientist working in the Middle East. While he found camels for sale for $100, the cost to get them to Indio ran upward of $3,000 each (nearly $30,000 today) because an attendant had to be sent and his return trip paid for. Additionally, the camels faced quarantine and a law that forbade breeding. In 1952, the fair spent $2,500 acquiring Jameel and Noor, a pair of dromedaries from Australia, where feral camel populations wreak havoc. (Both, courtesy of Linda Beal and the History Hunters.)

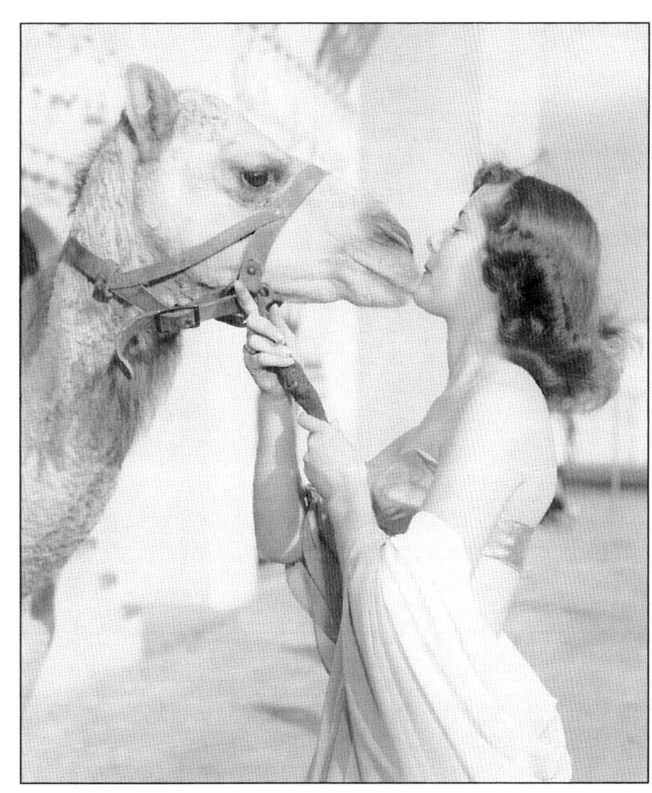

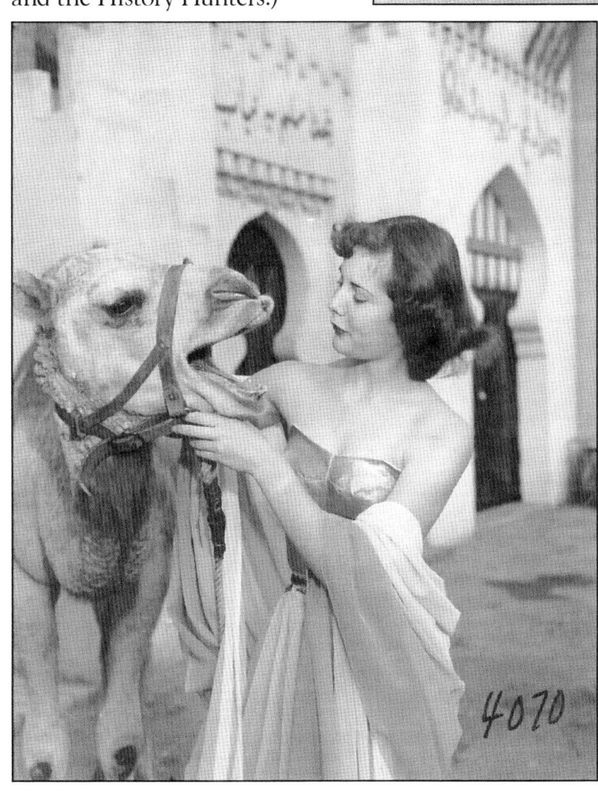

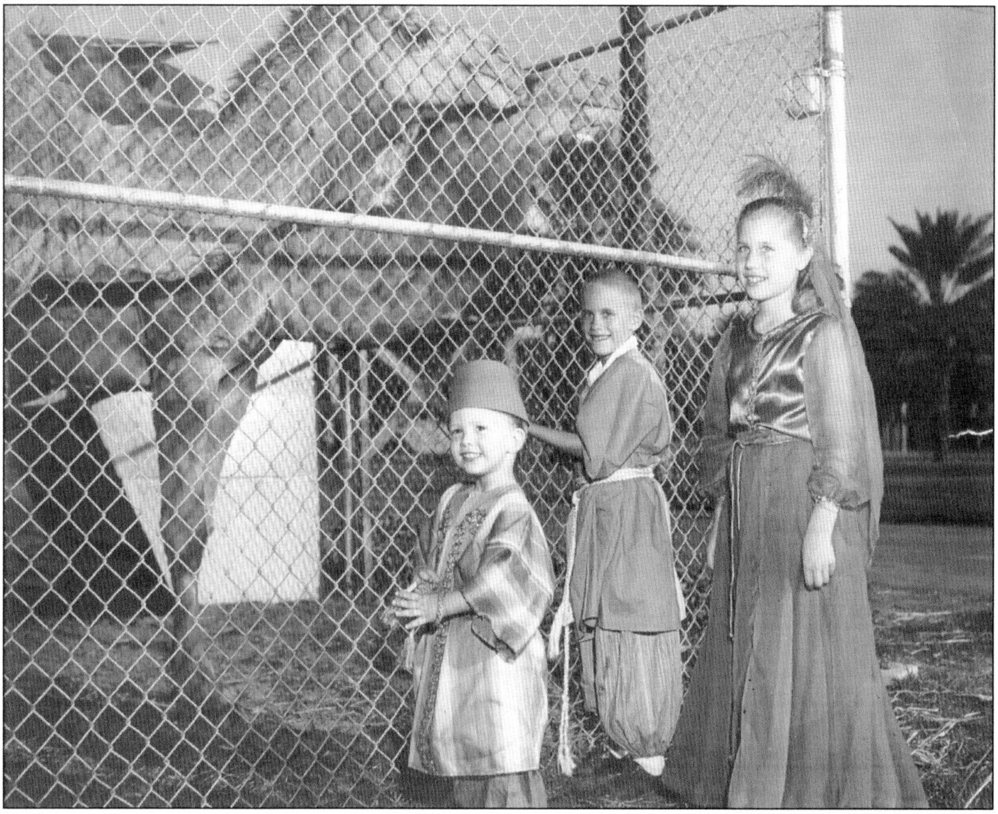

The camels earned the nicknames Jimmy and Nora, though in print the festival referred to them as Heba and Sheba to fit the theme more closely. Above, a trio stands near the camel pen in 1963. Below, in 1959, the fair traded Jamal for a young female camel born at an earlier fair. The festival held a children's contest to name the baby camel with the winner receiving a bicycle; Baladi, Arabic for "native girl," was chosen out of 2,000 submissions. Another naming contest was held in 1988 when baby Milagro was born during the festival. In 1990, a visiting camel gave birth to a baby girl, appropriately named Indio. The camel and mother pair pictured below attended an earlier festival. (Above, courtesy of the Coachella Valley History Museum; below, courtesy of Linda Beal and the History Hunters.)

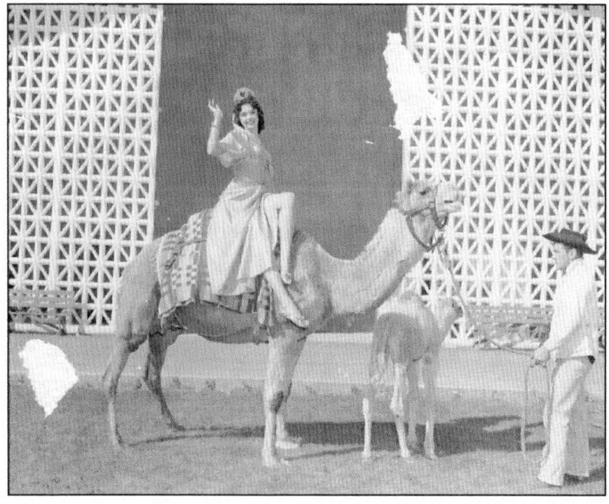

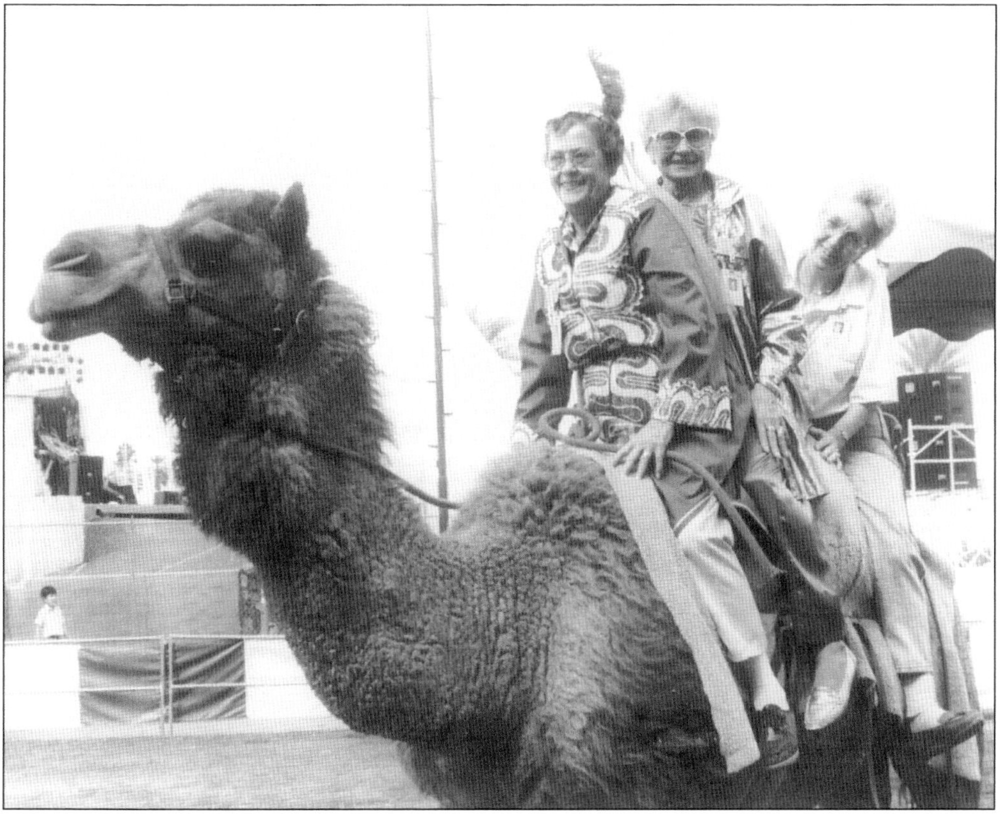

Riding camels remains a popular attraction. Dromedary camels, domesticated nearly 4,000 years ago, populate the Middle East and Northern Africa as beasts of burden. More than 90 percent of the world's camel population are dromedaries and transport goods and people. They can occasionally reach speeds of up to 40 miles per hour. Pictured from left to right, Alice Lowery, Jeri Taylor, and Joanne King sit astride this camel.

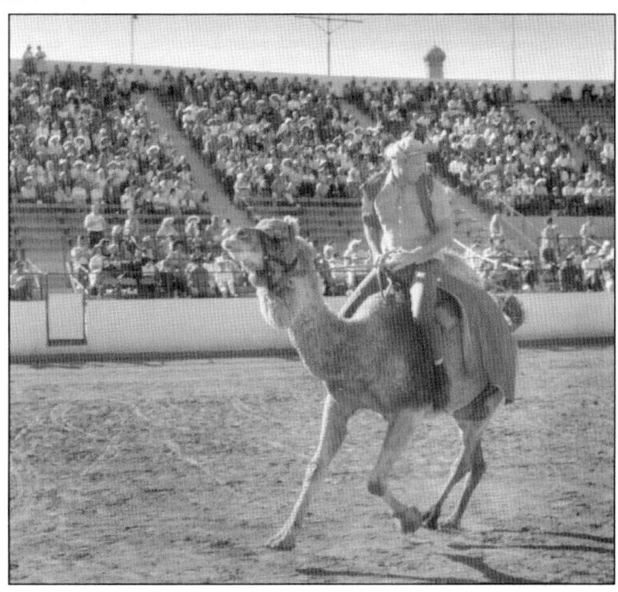

While the first public camel races occurred at the date festival, they were popularized in Virginia City, Nevada, as well. In fact, Virginia City claims to have originated the camel race in 1866 with animals that transported salt to the Comstock mines. They would not reappear in the town until 1959. In 1966, the date festival held a Virginia City Day. Today, the races share the camels, provided by Hedrick's Exotic Animal Farm.

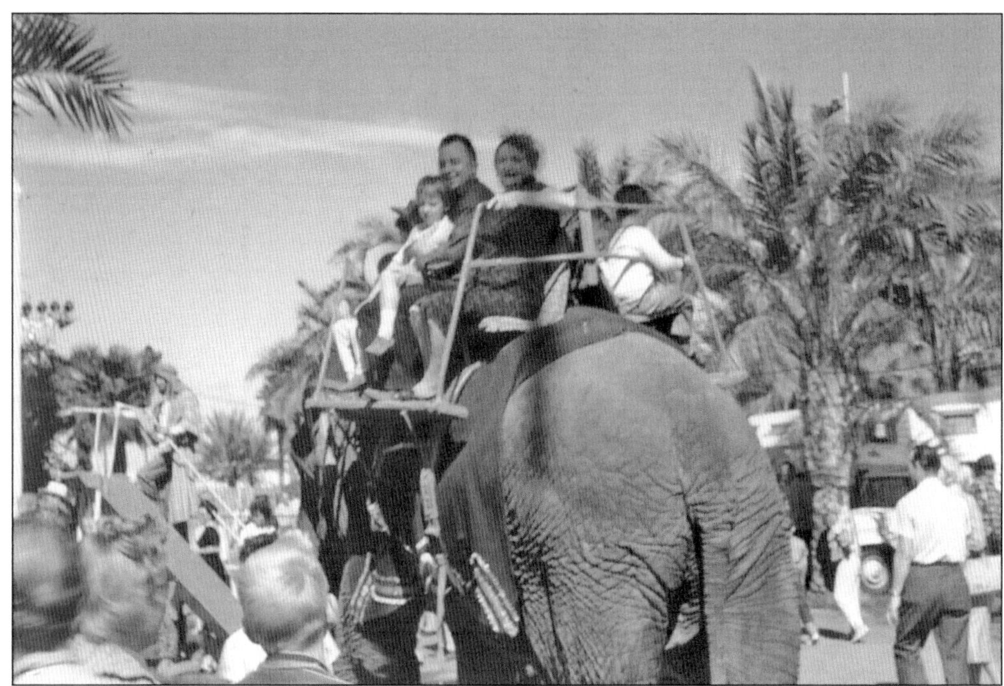

Elephant rides were also popular at the fair. Surprisingly, the elephant rides took place outside a corral, among the general fair-going public. The saddle allowed for six or more riders to sit atop the beast. Above, pictured from left to right, John Gault, Wendy Gault, and Ana Asker enjoy the ride about 1961. Elephants are not native to the Middle East. One species lives in sub-Saharan Africa while the other resides in Southeast Asia. Below, the author and her mother, Margo Ligman McCormick, enjoy an elephant ride in 1985. By this time, the elephants had been relegated to a corral. (Above, courtesy of the Asker family; below, photograph by Patrick McCormick, author's collection.)

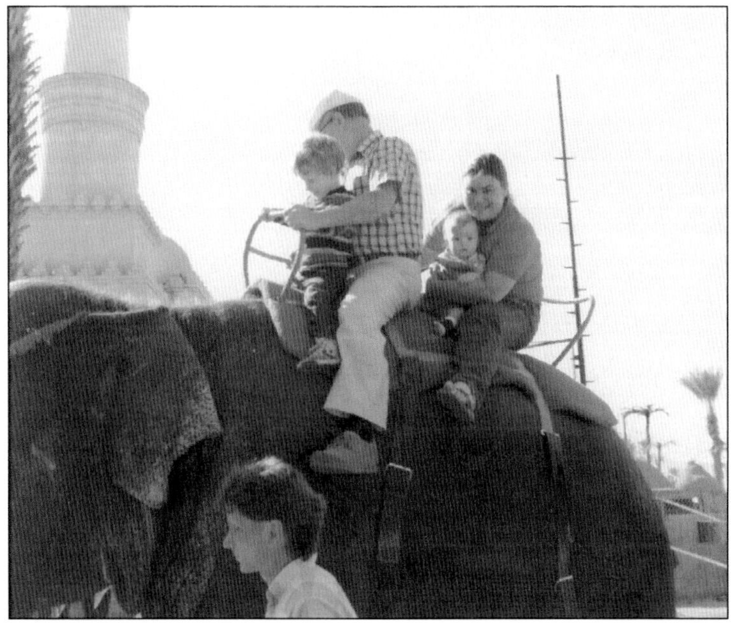

The elephants aided with their fair share of publicity, implying a general Orientalism but also reminding people of the fun associated with the circus, where elephants had long performed. A 1984 princess, Katherine Murray, poses with a trained elephant inside the fairgrounds.

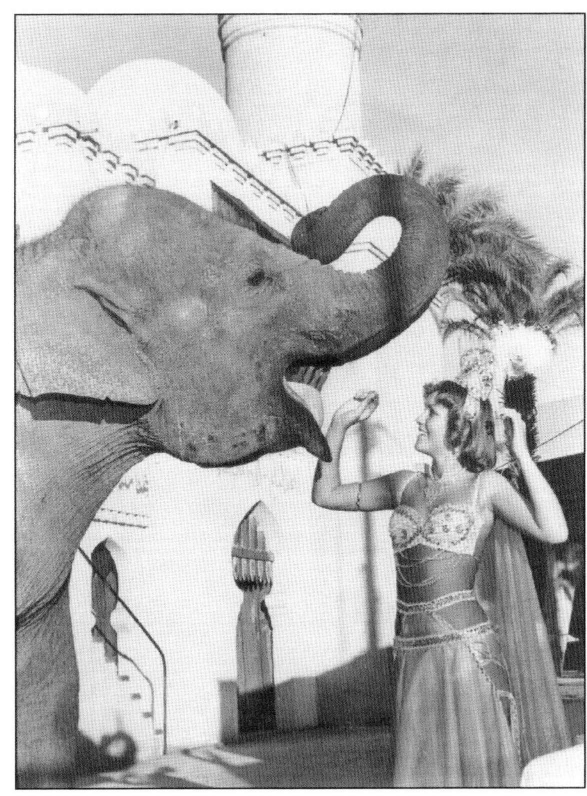

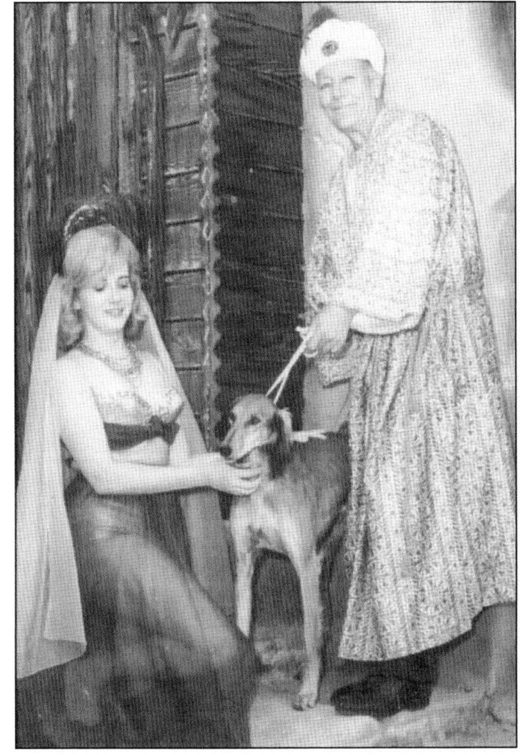

In 1966, the festival hosted a Saluki Walk, highlighting the oldest breed of dog. Referred to as the royal dog of Egypt, salukis have been found mummified alongside pharaohs. The Western Saluki Association dressed in Arabian-inspired garb and put their greyhound-like pets on display for the day. (Author's collection.)

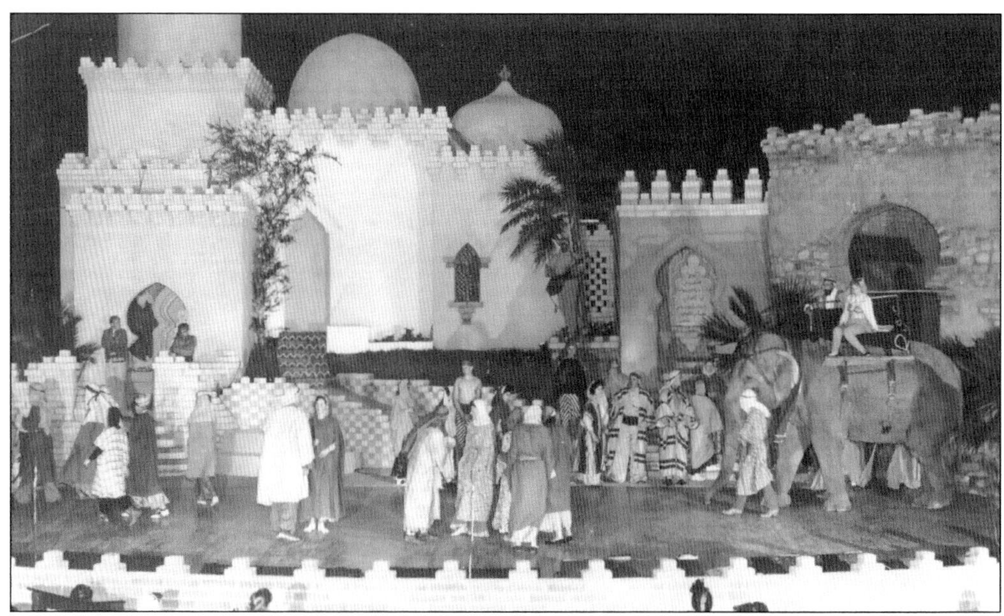

Animals held more than one job at the fair. In addition to racing and giving visitors rides, the camels often found roles onstage in the pageant. Not to be outdone, the elephants were paraded across the stage as well (above), as seen here in 1956. Of course the pageant also spotlighted other exotic animals, including these tiger cubs from the 1969 festival (below). While local talent made up most of the pageant's cast, the animals were always handled by professionals, who seldom had to wear such exotic costumes at the other festivals and events they attended. (Above, courtesy of the Coachella Valley History Museum; below, author's collection.)

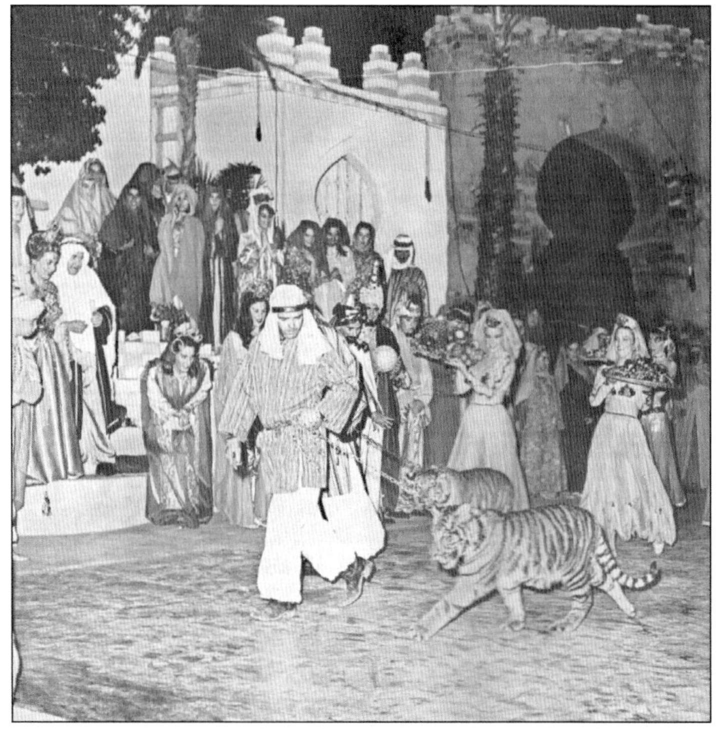

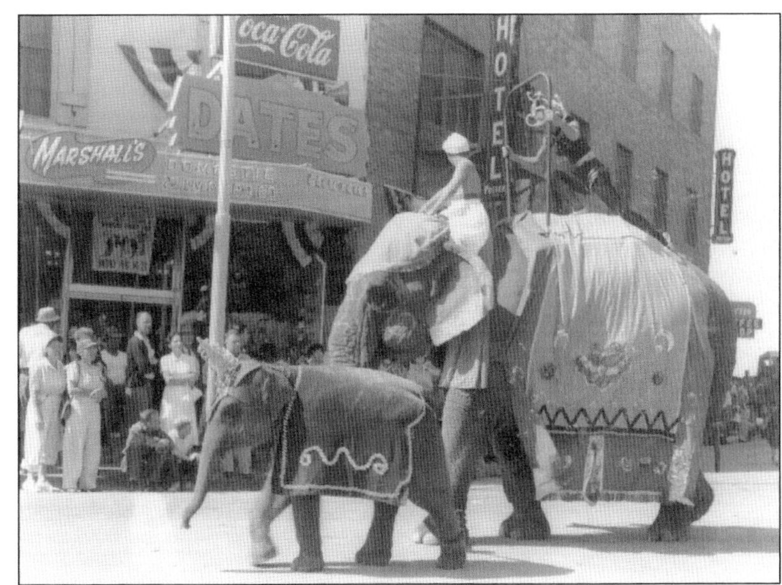

After providing rides, racing, or starring in the pageant, several of the animals made appearances at the annual parade. Elephants drew excited cheers, and camels reminded parade watchers of the festival's theme. The animals even wore costumes, as seen here by the elephant pair with face paint and headdresses.

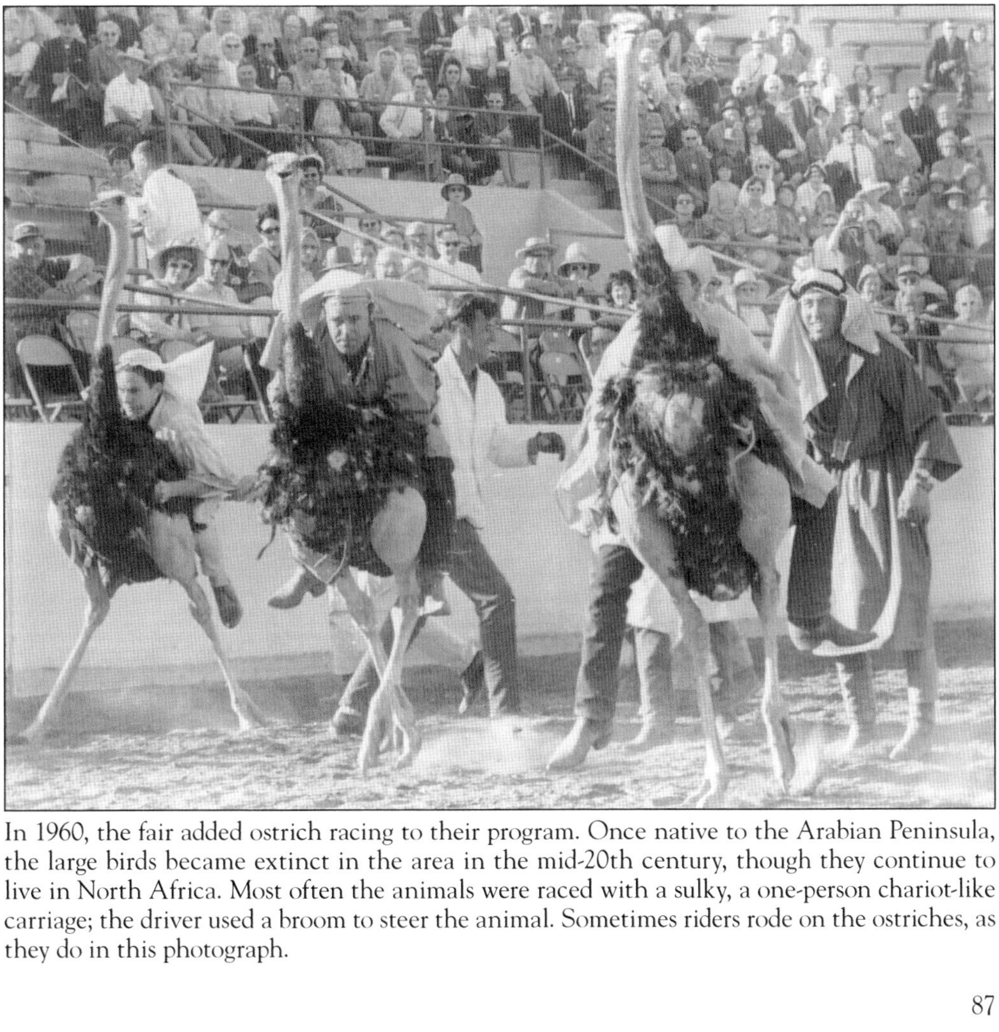

In 1960, the fair added ostrich racing to their program. Once native to the Arabian Peninsula, the large birds became extinct in the area in the mid-20th century, though they continue to live in North Africa. Most often the animals were raced with a sulky, a one-person chariot-like carriage; the driver used a broom to steer the animal. Sometimes riders rode on the ostriches, as they do in this photograph.

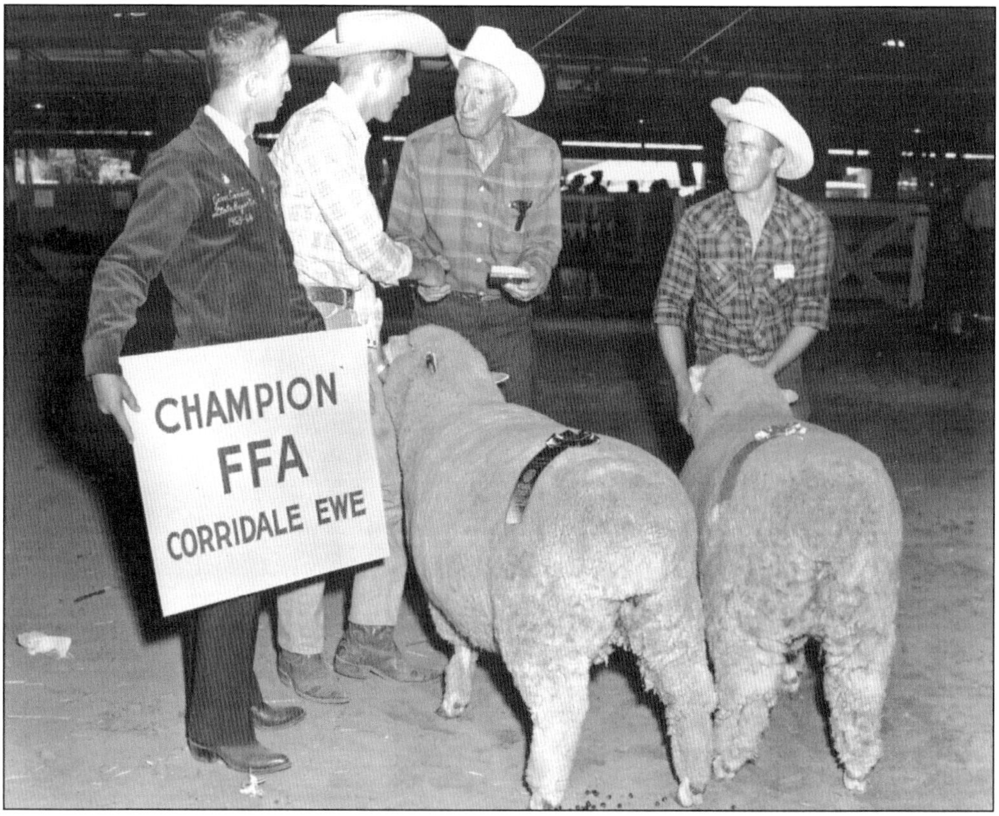

Even today, youth from around the county raise livestock to show at the date festival. Future Farmers of America and 4-H members bring horses, cows, sheep, goats, pigs, and fowl to the fair. Best specimens win awards, and students often sell the animals afterward to recoup some of the costs. Although the Future Farmers of America incorporated in 1928, young women were denied membership until 1969. Above, the California state reporter for the Future Farmers of America holds a sign indicating the winning ewe for 1964. Tied to the US Department of Agriculture, 4-H encourages youth to embrace scientific advancement and farming technology. Below, a 4-H participant poses with his prizewinning lamb around 1982. (Above, courtesy of the Coachella Valley Water District; below, author's collection.)

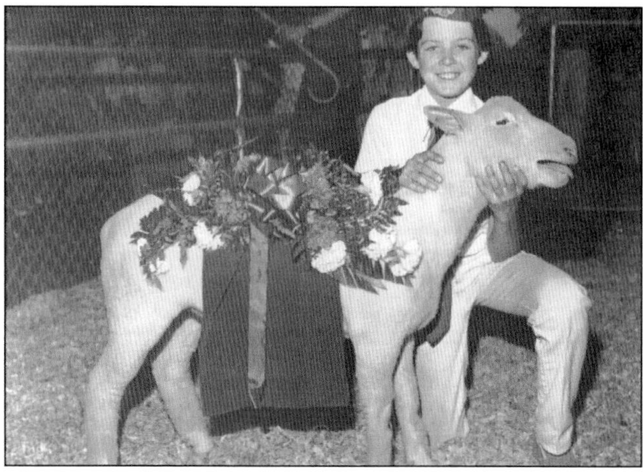

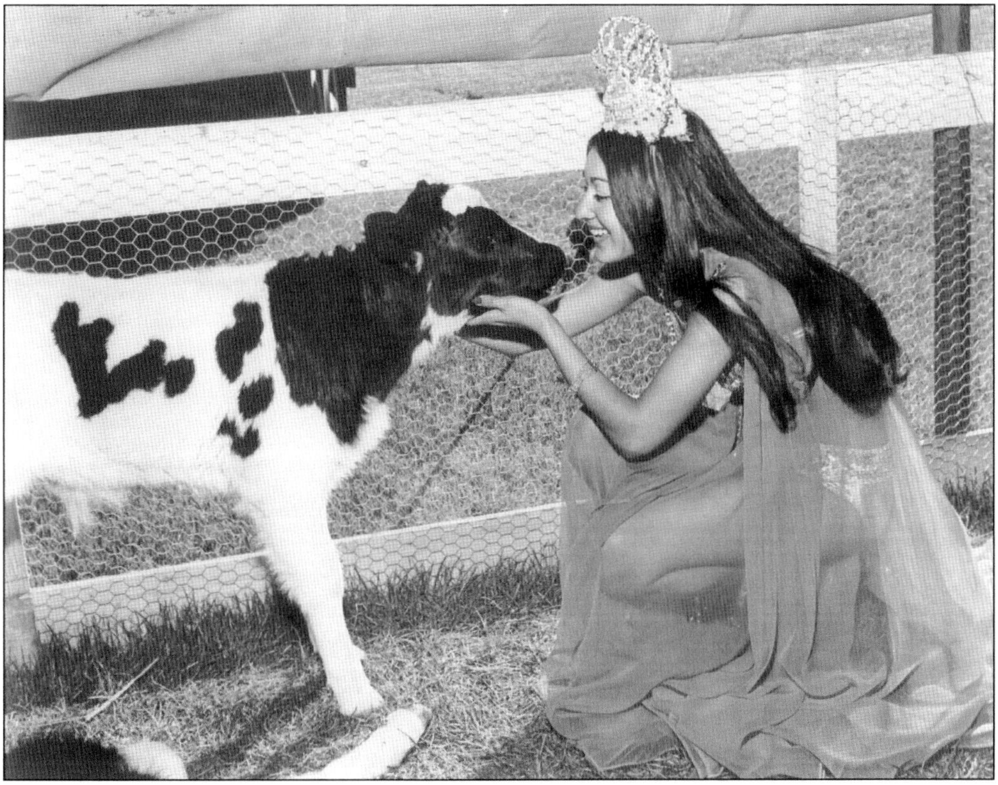

Though livestock served as a means for youth to learn the ins and outs of agriculture, it was also a draw for fair visitors. As the region lost its family farms, the date festival remained a place where children and adults could interact with animals typically found on the farm. Rita Pescador, Miss Coachella, plays with a young calf in this 1972 photograph.

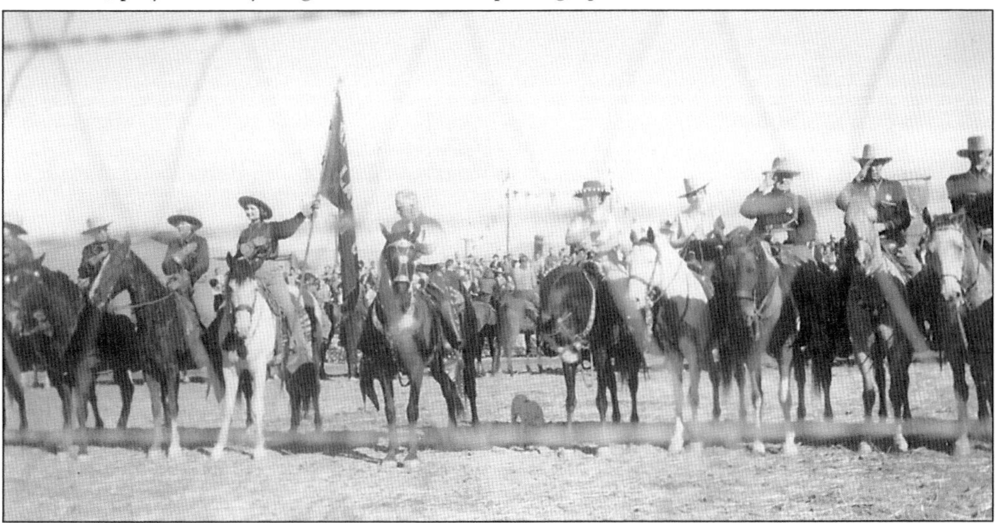

Before the Riverside County Fair joined the date festival in Indio, the city held an annual stampede rodeo. Riders traveled from town to town to participate in similar events, with Palm Springs hosting a rodeo around the same time. Events drew large crowds and inspired the Western theme common to early fairs.

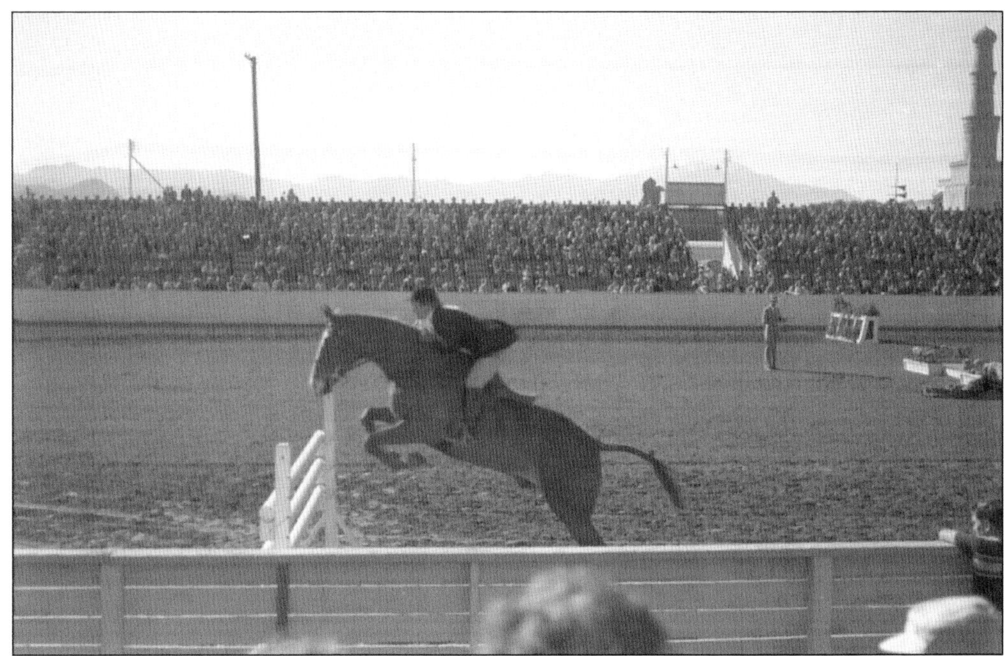

In addition to the rodeo, a national horseshow was added to the event schedule. Participants earned cash prizes and trophies for jumping horses, carriage-driving ponies, and parade horses. By 1959, more than 500 horses competed. Events were scheduled every afternoon, with a special Mexican *charreadas* held occasionally. Above, a large audience looks on as a participant soars across a jump. Below, Martha Castro presents a ribbon and decanter trophy to an entrant in a sulky competition. (Above, photograph by Glenn Stocking, courtesy of the Schleis family; below, courtesy of the Castro family.)

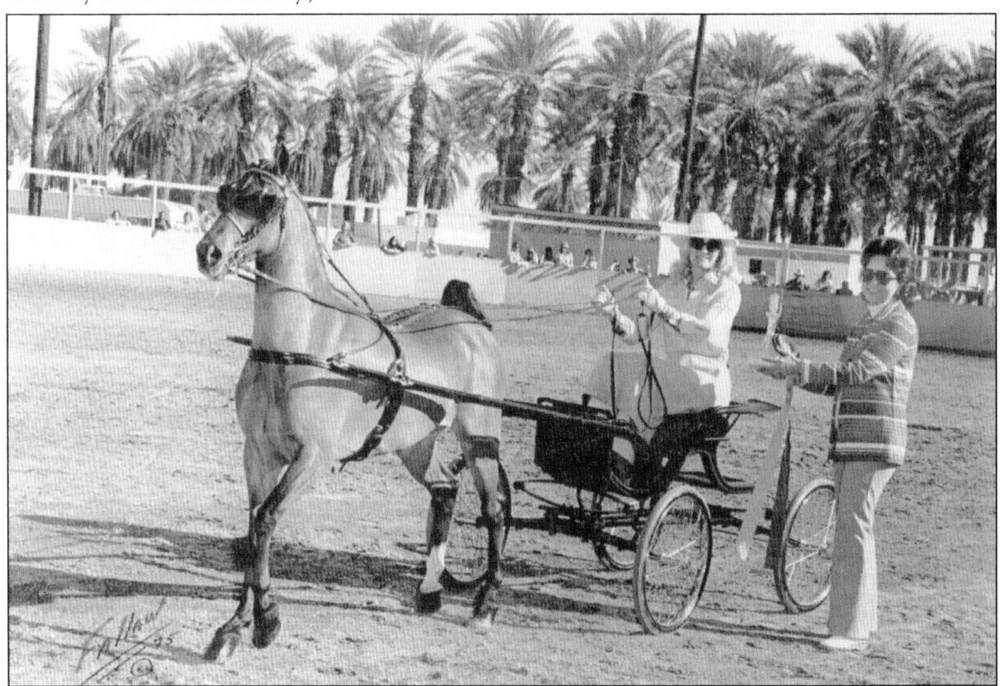

Six

Waving Genies
The Annual Parade

Held annually on Presidents Day, the Riverside County Fair and National Date Festival Parade continues to draw attention today. Early parades embraced the fair's rodeo, championing Western wear and mounted entries. After the Arabian theme was chosen, floats bore genies, harem girls, and sultans, not to mention date palms. Camels and elephants paraded through Indio, while bands from local schools provided music for the revelry. The original route followed a similar path as today's parade, down Miles Avenue, then across downtown Indio toward the fairgrounds on today's Highway 111 (then Avenue 46). It not only highlighted Indio's business center but also passed by several date palms, further connecting the Arabian-themed celebration to its agricultural origin.

During the fair's heyday, community support overflowed. Local businesses entered floats for publicity, and service clubs did the same, hoping to boost community spirit. Newspapers commented that nearly the entire town of Indio, plus hundreds of visitors, attended the event. It was so important to the city that the mayor issued proclamations requesting that all businesses close during the parade so that employees could attend. With so many people standing along the route, it is likely the businesses would not have had many customers, anyway. The best floats in the service club, civic and fraternal organization, and commercial categories received trophies, as did various mounted entries. One year, 1950, even saw the addition of a night parade, complete with lighted floats.

In order to draw attention to the festival, fair organizers entered floats into other regional parades. Queen Scheherazade contestants often rode in style through the streets of other towns, encouraging the audience to visit the date festival. In 1957, for example, Indio's festival float won sweepstakes at the 68th Tournament of Roses Parade. Reports indicated this publicity stunt introduced Indio to people as far away as London.

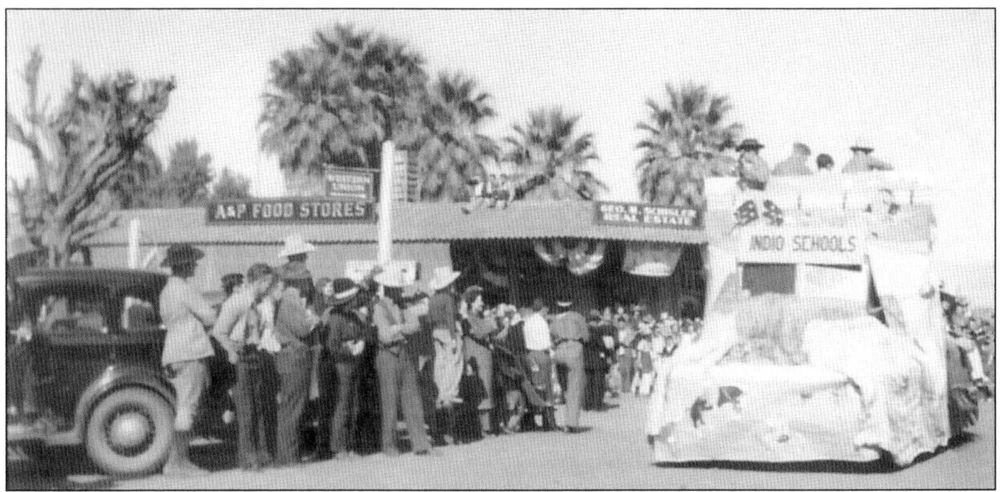

The Western theme shines through in this pre-1941 fair float, entered by the Indio schools. Hand-painted scenes of a rodeo, with bulls and riders, decorate the front of the truck, while Indio's schoolchildren, in cowboy hats and boots, ride in the back.

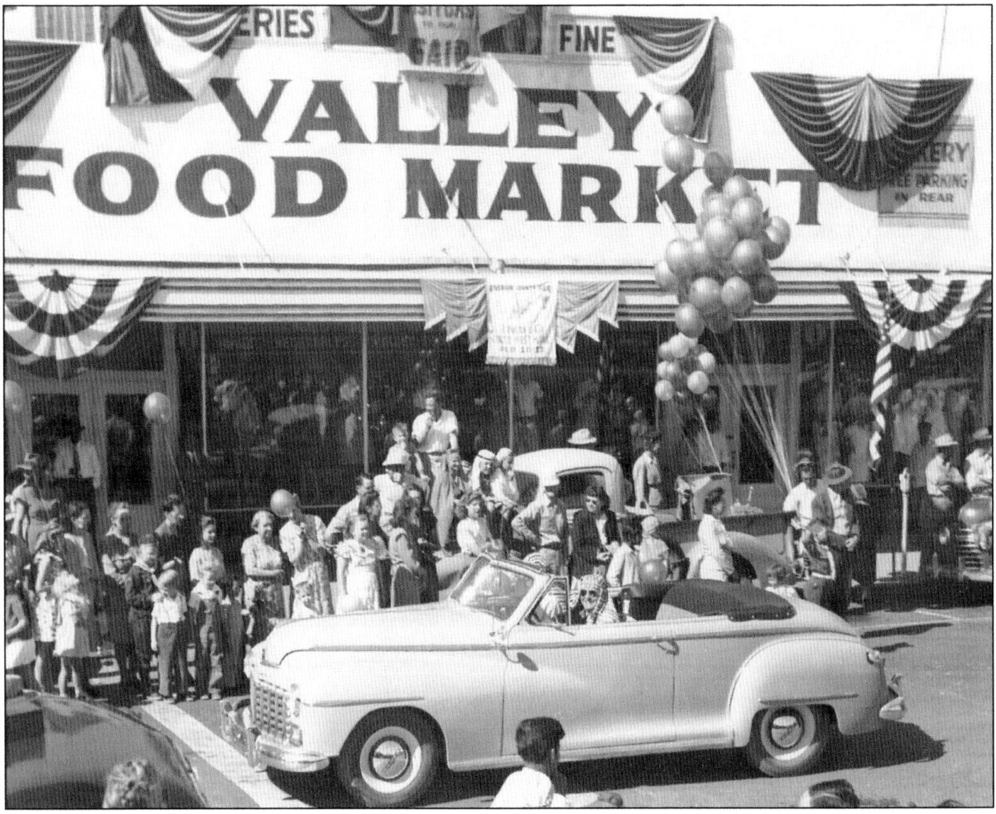

With so many out-of-town visitors attending the fair and parade, Indio wanted to put its best foot forward. Service clubs organized clean-up days around the city, and the mayor asked people to ensure their property was clear and presentable. Organizers also hired a decorating company out of Fresno, California. Notice the plethora of banners, balloons, and decorations in this 1950s photograph, made possible through the cooperation of local merchants. (Photograph by Bradshaw.)

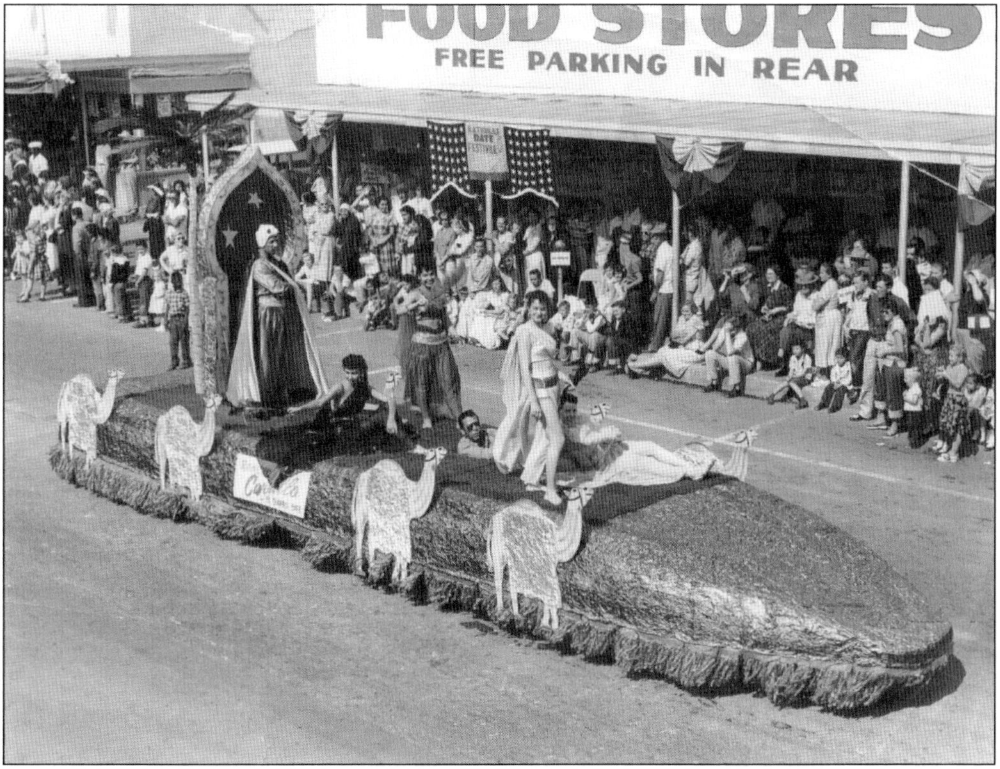

Beloved by Indio residents, the Rancho Carrillo and El Charro Café served food prepared by owner Jess Carrillo. While the top photograph indicates his embrace of the Middle Eastern theme, the bottom reflects the date festival's inclusion of the Mexican American community. On the Arabian float from 1955 are sultan Roy Rodarte, Dora Demarbiex (sitting), Mary Carrillo Carmona (standing), and Sandra Brooks Lafayette (standing front). On the Mexican-style float from the late 1940s, Carillo's daughter Mary Carrillo Carmona stands to the right of an unidentified Mexican singer who entertained guests at the restaurants. Polly Redwine stands to her left, while the Dos Reales musical group plays guitars behind them. (Photographs by Lucille Stewart, courtesy of Mary Carrillo Carmona.)

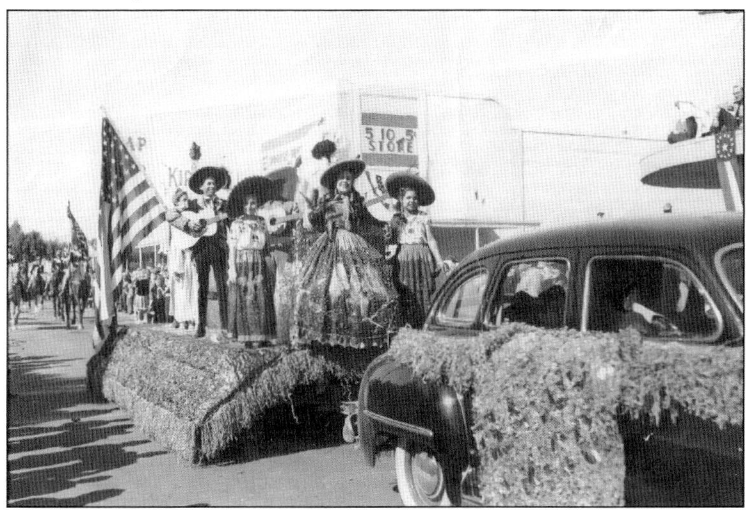

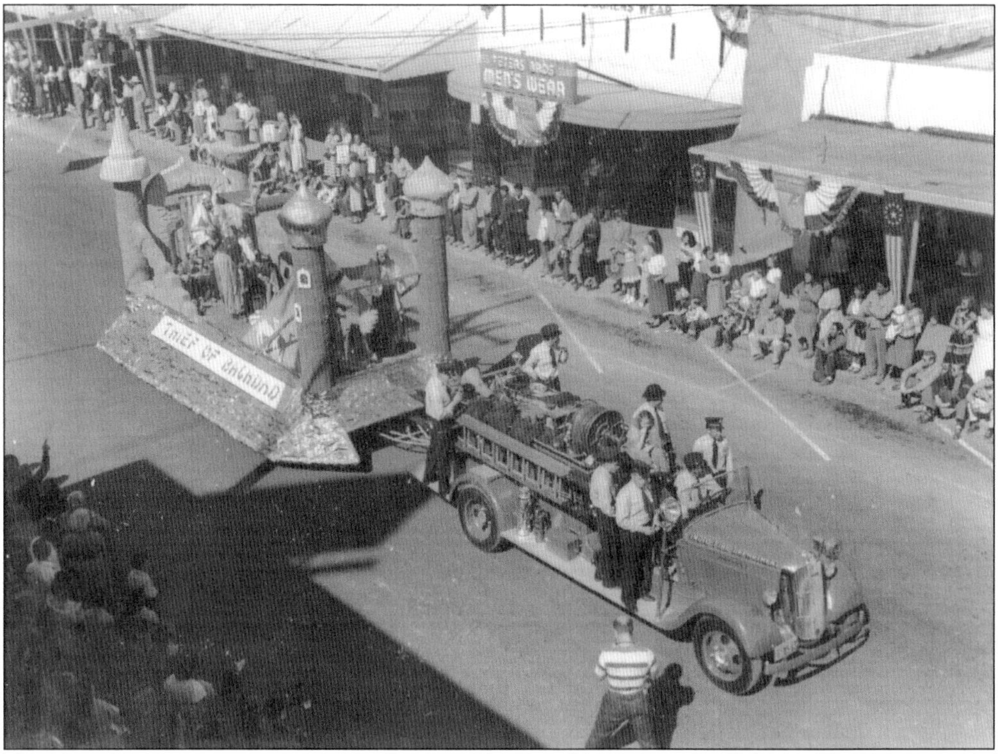

Indio's firefighters not only rescued residents but also pulled floats in the parade. Above, the engine tows a float depicting scenes from the *Thief of Bagdad*. The 1924 film and its 1940 remake loosely mirrored *One Thousand and One Arabian Nights*. The Peters Brothers Clothing Store, visible in the photograph, was owned by the Peters family, emigrants from Syria, some of the few Indio residents of Middle Eastern descent. Below, firemen recreate a scene from 1924 in the parade. Hand-drawn hose carts like this one would have been carried to a fire by the firefighters themselves. Since the position was voluntary at the time, the men would have rushed from their homes, hence the pajama costumes worn by the 1954 parade participants.

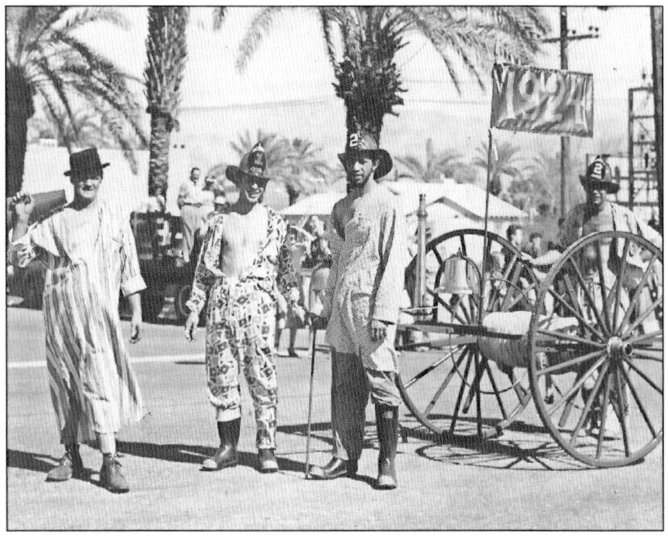

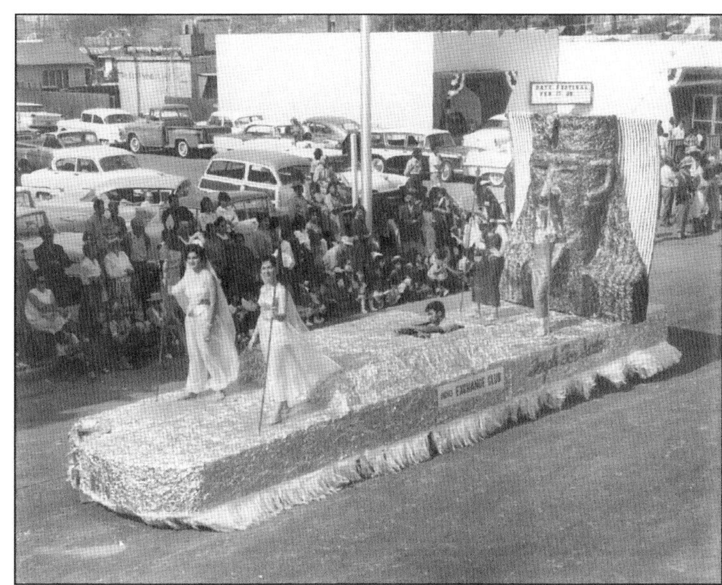

During Indio's attempt to bring a Nubian temple to the Coachella Valley (see page 46), the Exchange Club created a float to fit the cause. Lynne Marie Roman rides on the front of the 1961 Egyptian-themed float on the corner of Jackson/Fargo and Wilson Streets. Indio Cleaners, located behind the float, advertised the date festival dates on their marquee. (Courtesy of Lynne Marie Roman Perry.)

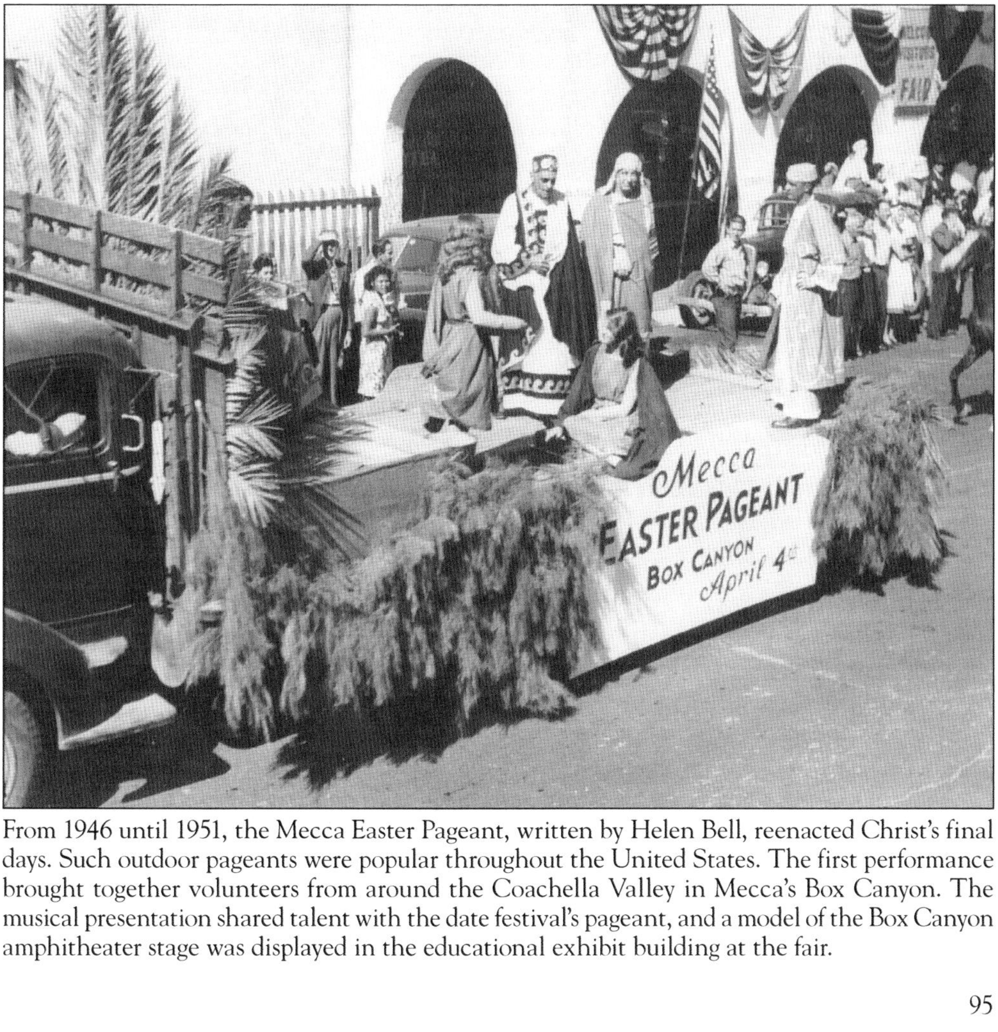

From 1946 until 1951, the Mecca Easter Pageant, written by Helen Bell, reenacted Christ's final days. Such outdoor pageants were popular throughout the United States. The first performance brought together volunteers from around the Coachella Valley in Mecca's Box Canyon. The musical presentation shared talent with the date festival's pageant, and a model of the Box Canyon amphitheater stage was displayed in the educational exhibit building at the fair.

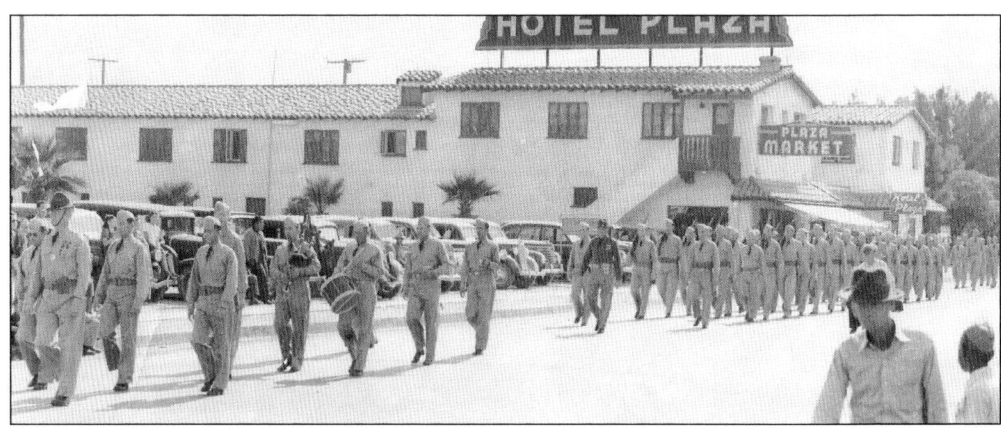

Indio hosted several veterans' associations, including posts of the American Legion and Veterans of Foreign Wars. During World War II, the nearby Desert Training Center educated over one million men, who sometimes relocated to the area after they returned from war. Current and former members of the military marched in the parade annually. Above, a group associated with the armed forces, possibly a veterans' organization, passes the famed Hotel Plaza. Below, J.C. Rose, district inspector during World War I, rides on the Veterans of World War I float. The organization united servicemen beginning in 1948, roughly 30 years after their service.

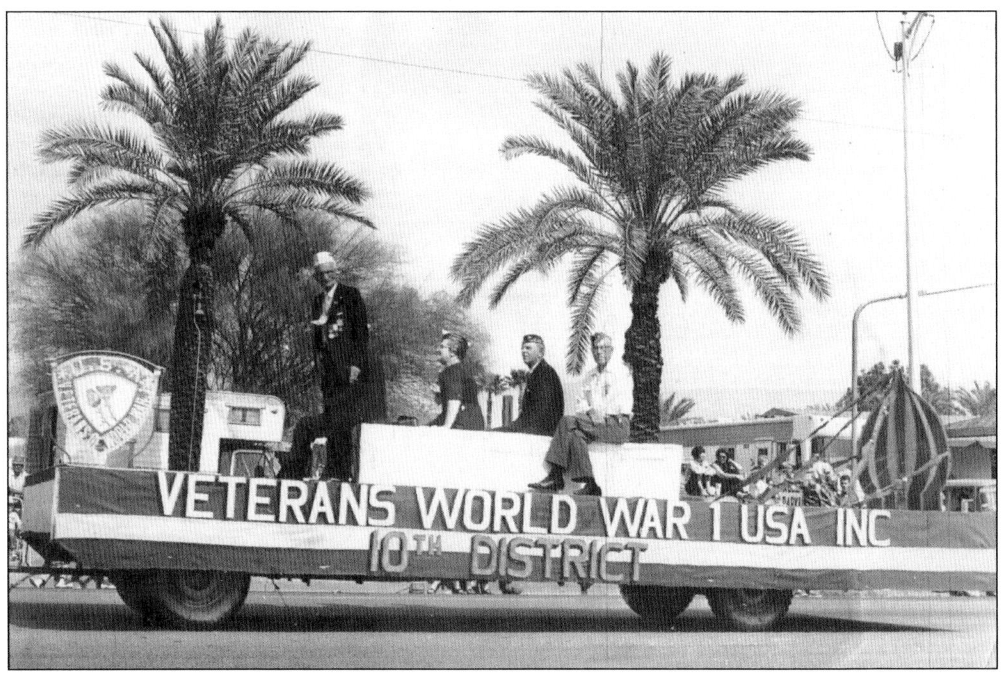

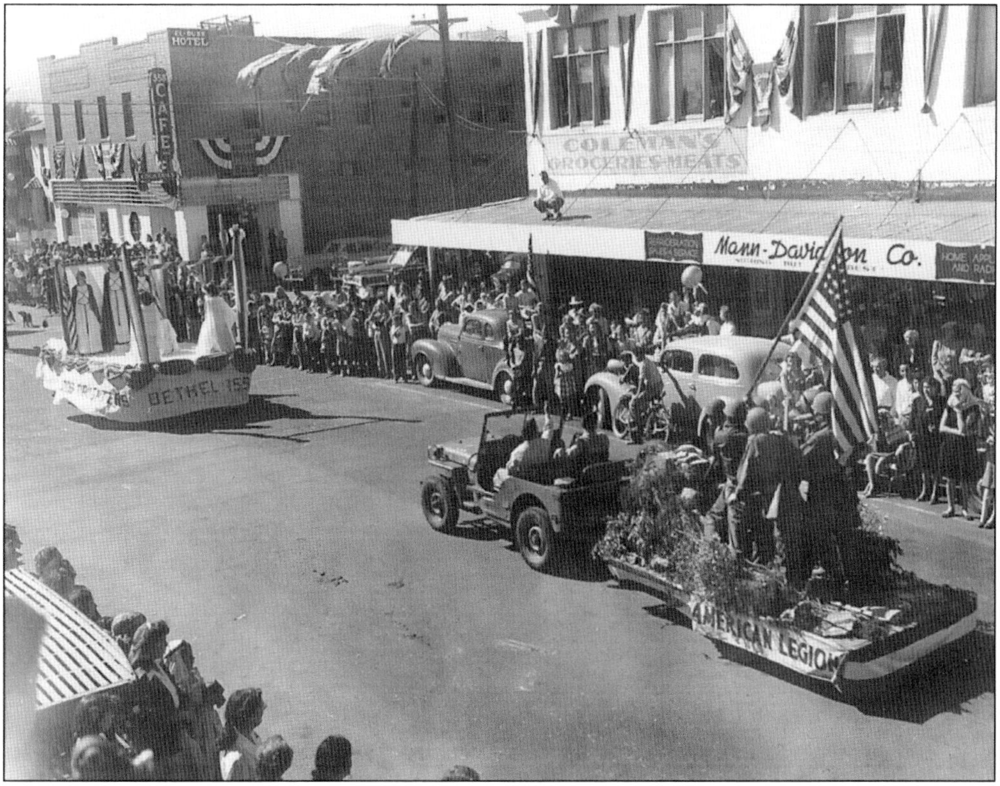

American Legion posts around the county sponsored the early queen pageants, overseeing voting in each community before the local winner went on to compete for the larger county title. The Coachella Valley post reenacts the famous scene at Iwo Jima for their 1947 float. An army-grade jeep tows the entry.

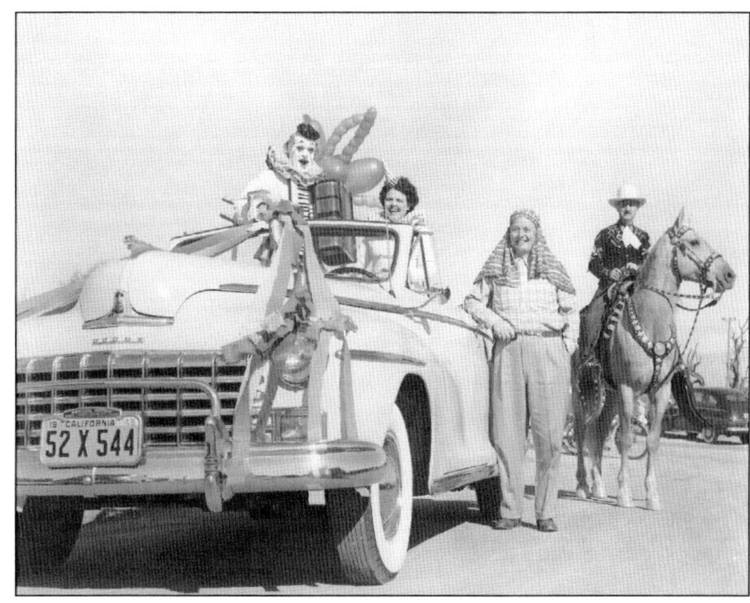

Clowns were a crowd favorite, their joy contagious even in this 1948 photograph. While the clowns brought color to the parade, even the cars were decorated. Crepe paper and balloons adorn an Imperial Motors Dodge car, while a mounted entry looks on. (Photograph by Bradshaw.)

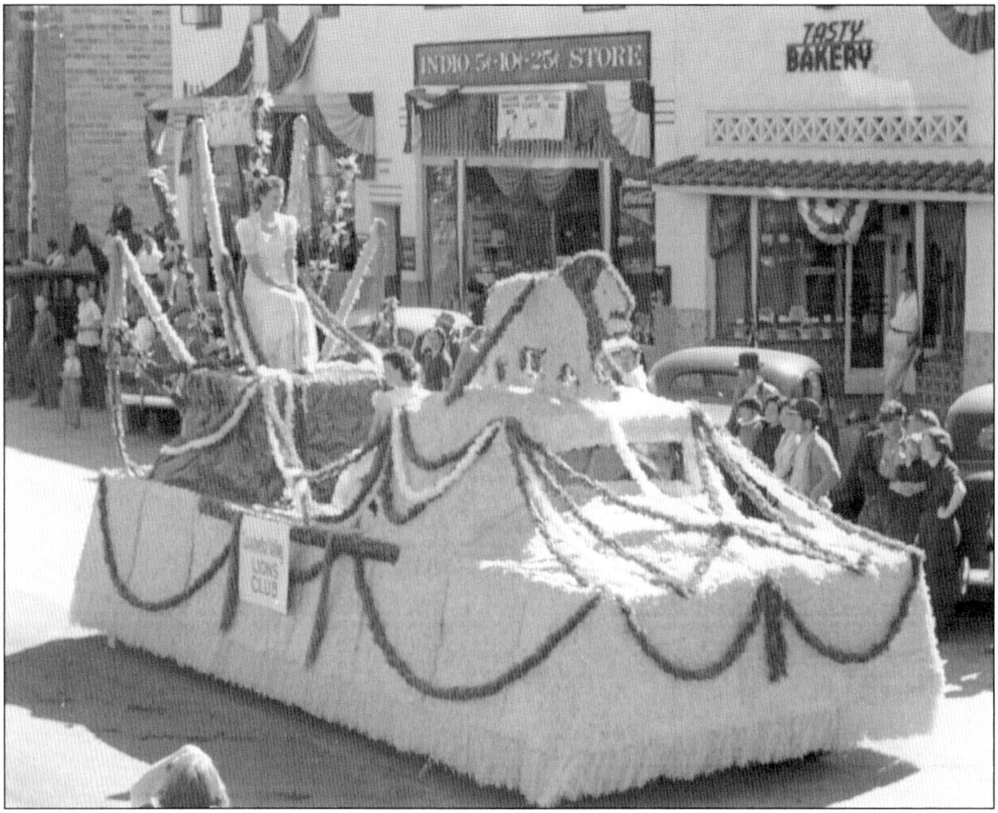

Still active today, the Coachella Valley Lions Club works on service projects throughout the region, including those aimed at preserving eyesight. They offered a float to the parade annually, often incorporating a lion. The Lions hosted a popular pancake breakfast before the parade to raise money for their service projects. Above, Margaret Proctor Tyler, wife of Lions member and dentist Dr. John Tyler, rides atop their 1940 float. Below, this entry includes both the lion and the Arabian theme. Notice the additional lion in the cage is one of the club members. (Above, photograph by Dal Woodhouse, courtesy of Margaret Tyler; below, courtesy of the Coachella Valley History Museum.)

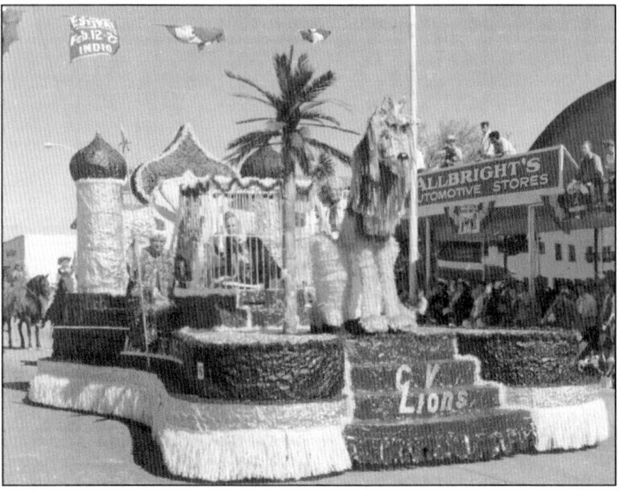

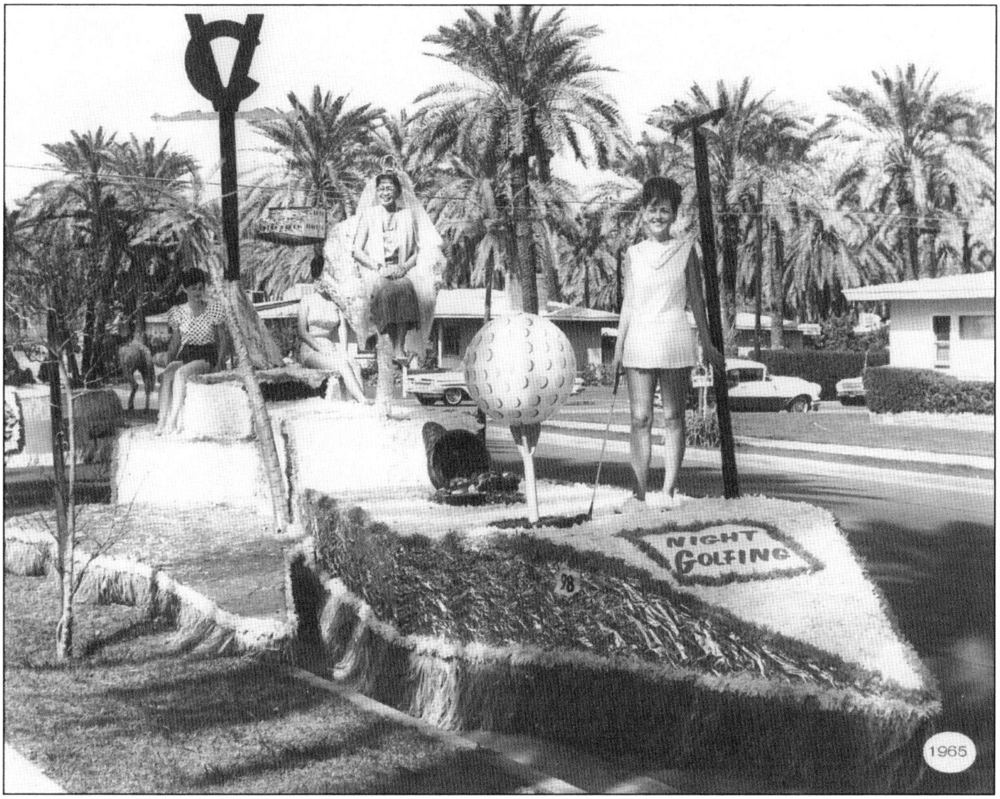

The City of Indio constructed one of the few nighttime golfing venues in the Coachella Valley, operating on the municipal golf course. The sport drew tourists not only to Palm Springs but also surrounding desert communities and proved popular with locals as well. The 1965 night golf float is above par. The float features Zonnie Wah (seated) and Beatrice Moreno (standing).

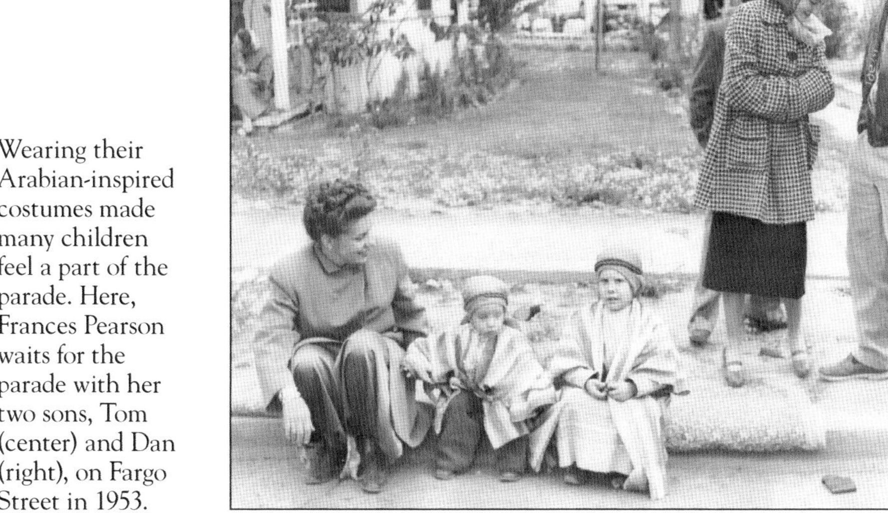

Wearing their Arabian-inspired costumes made many children feel a part of the parade. Here, Frances Pearson waits for the parade with her two sons, Tom (center) and Dan (right), on Fargo Street in 1953.

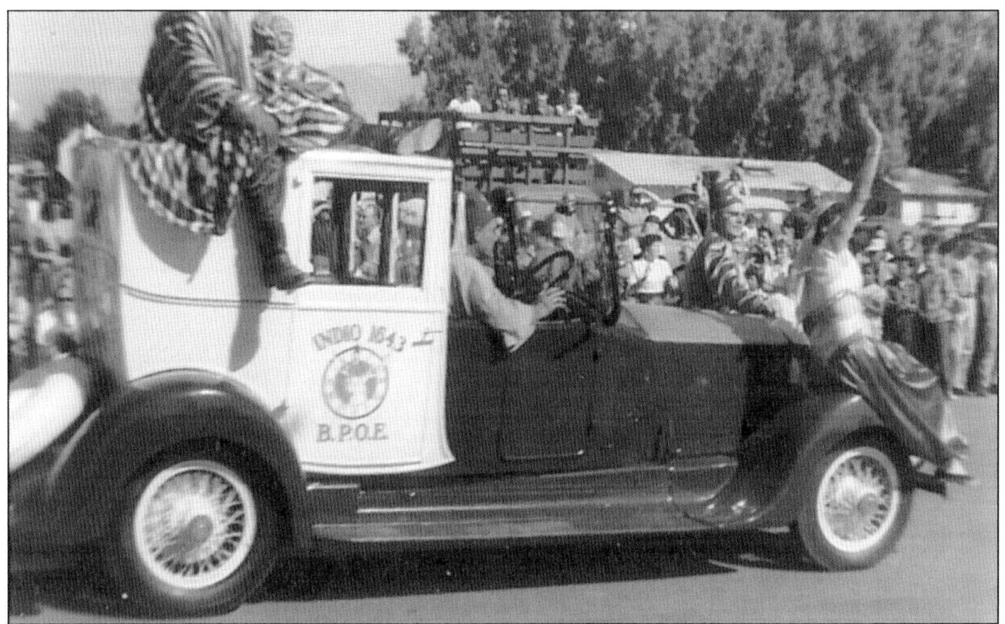

Fraternal organizations participated at every level of the festival, including the parade. Members of the Benevolent and Protective Order of Elks drove their car through the streets of Indio. Women were unable to join until the 1990s, so the woman sitting on the car may have belonged to one of the unofficial auxiliary clubs. Traditionally, the Elks aid veterans' services and youth programs around the nation.

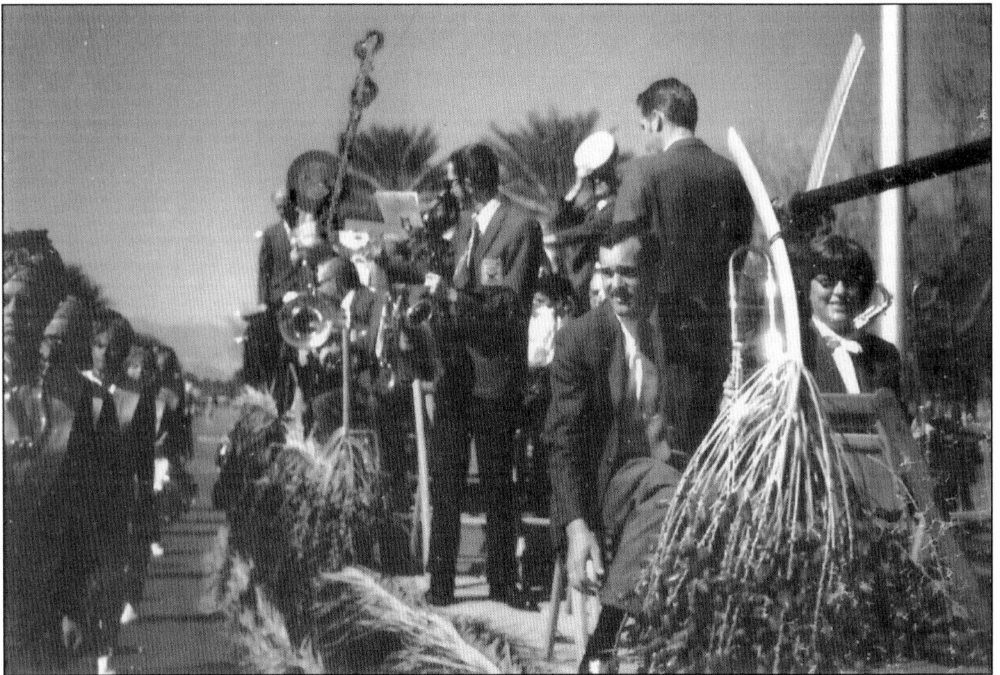

The Indio Municipal Band brought music and merriment to the parades. The band took the occasion quite literally and decorated their float with palm fronds and clusters of date fruit. To the left of the float, the Indio Rajah Marching Band shows off their rajah-inspired uniforms.

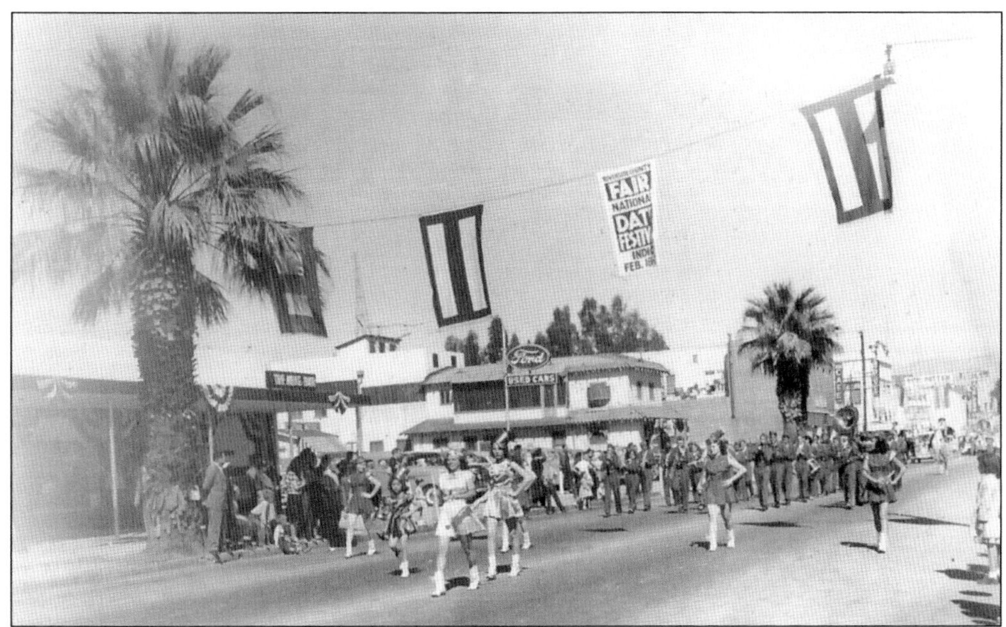

School bands provided students with a prime place to share their love of music. Above, an unidentified band marches along the parade route. The Hotel Potter and the Elk's Lodge can be seen behind them. Below, Coachella Valley Union High School, whose mascot has long been the Arabs, participated in the parade annually. Built in 1910, it served all Eastern Valley students until Indio High School opened in 1958. Since so many parade watchers were alumni, the Arab Marching Band proved a fan favorite. Here, the 1953 band readies for a performance. Though they resembled date festival costumes, the band wore their unique uniforms at most performances. (Above, courtesy of the Coachella Valley History Museum; below, author's collection.)

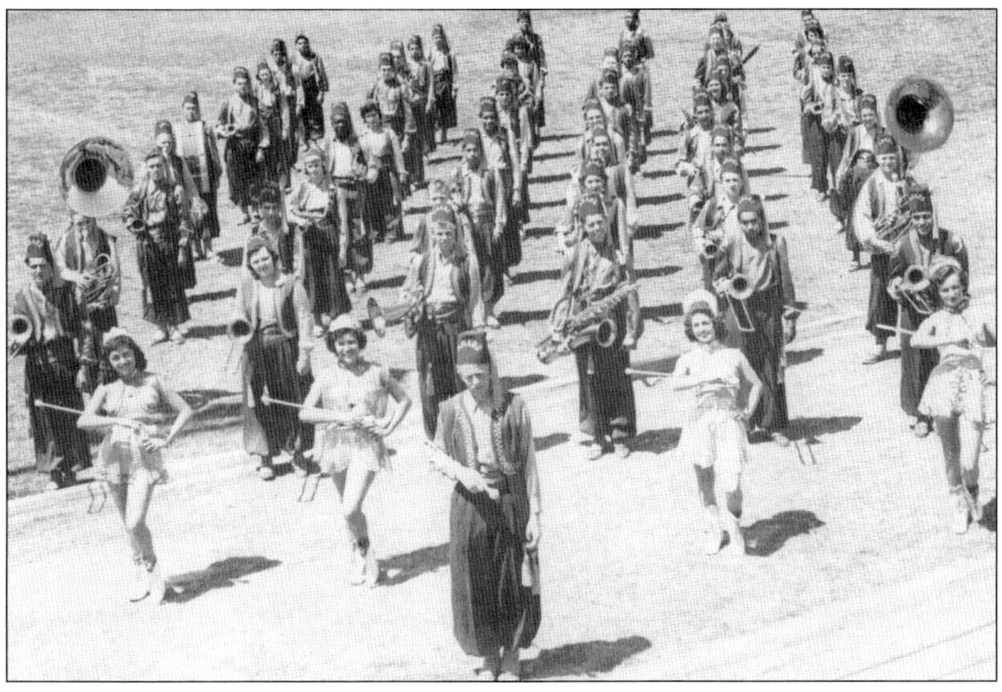

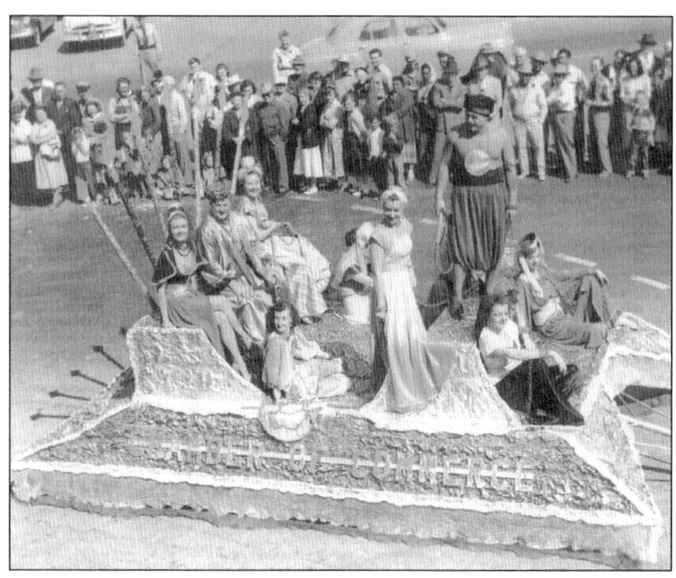

Organized in 1948, the Indio Chamber of Commerce serves as one of the community's biggest boosters. Today, the organization sponsors events such as the Southwest Arts Festival and encourages tourism to the area. Note the women in chains on this float designed to look like an Arabian slave market. Slavery was long associated with the harem in American popular culture. Allan Willard served as the sultan while Frank Rivers played the slave driver.

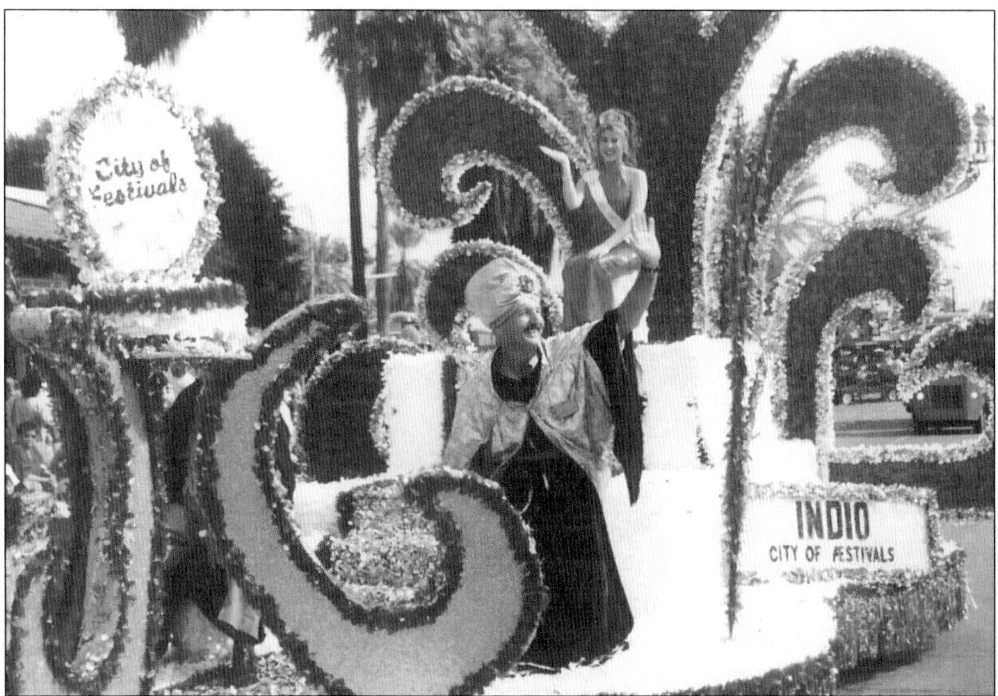

Though a vibrant community beginning in the early 20th century, the city of Indio was unincorporated until May 1930. Eventually, the city birthed other festivals: a well-known tamale festival, the Southwest Arts Festival, and, in more recent years, the Coachella and Stagecoach music festivals, earning the town the nickname the "City of Festivals." Doug York, who served on the city council, rides on the 1999 float advertising the moniker.

Audience members searched for the best seats in the house, even if that led to some unconventional locations. In addition to lining the parade route, residents took to the rooftops to see this 1957 Rotary International float. The service club was celebrating the 50th anniversary of Rotary's founding in Chicago, Illinois. (Courtesy of Desert Hot Springs Rotary.)

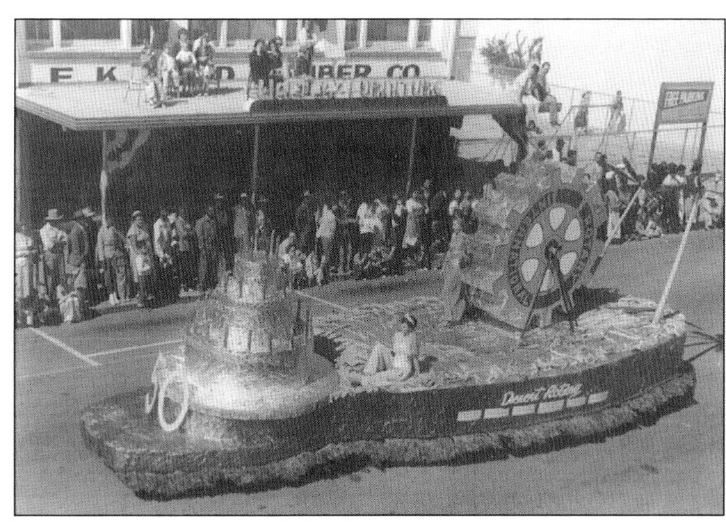

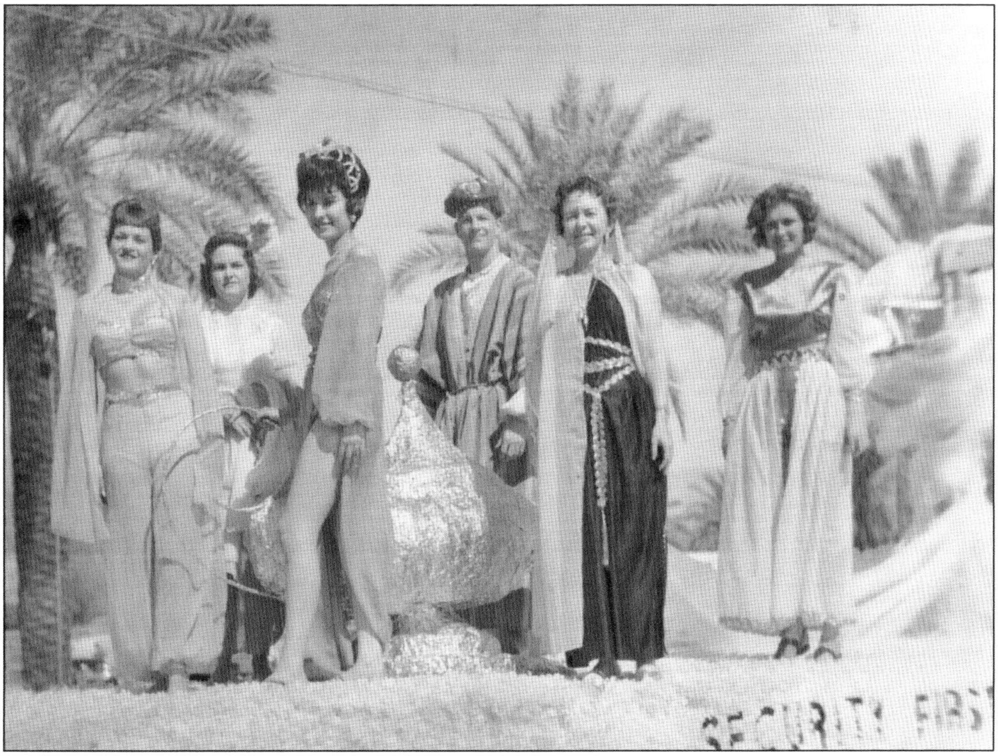

Employees from local businesses supported the fair in numerous ways. In addition to wearing costumes during fair time, many worked on the floats that traversed Miles Avenue every year. Security First Bank employees, for example, rode on a float annually. (Photograph by Field.)

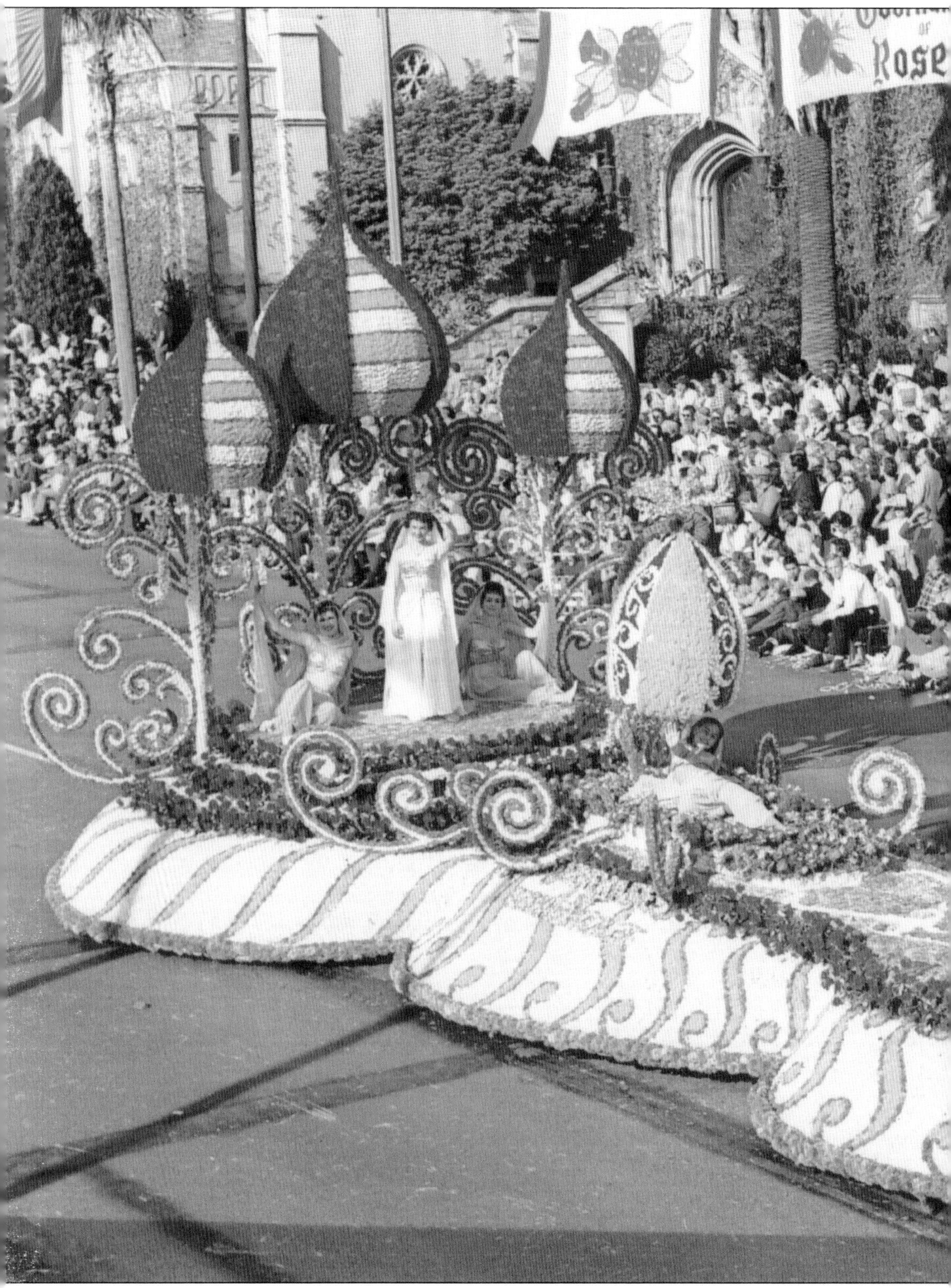

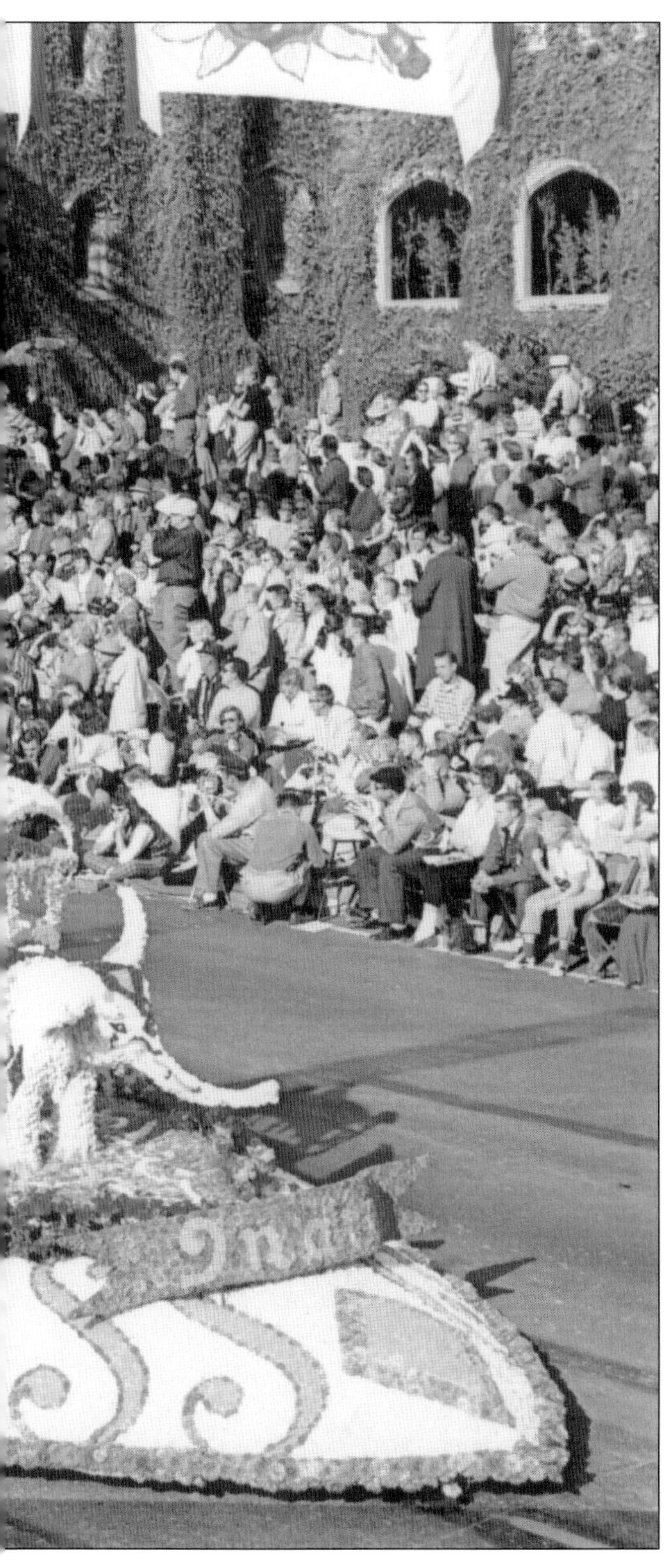

Indio residents hoped to draw attention (and visitors) to their city by entering a float into the 68th annual Tournament of Roses Parade in 1957 with the theme "Famous First in Flowers." To their delight, the float, "First Date Festival," won sweepstakes. Designed by Sam Coleman of Glendale, the float cost approximately $9,000 and featured 20,000 vanda orchids. To raise money, the Indio Chamber of Commerce sold bumper stickers that read, "I helped Indio join the Rose Parade" for $5. They also sponsored a competition for teenagers; the girls that sold the most flowers were eligible to ride on the float. The float depicted a caliph's garden. To enrich the theme, two camels preceded the float. The following year, the city entered "Jewels of the Magi" into the parade. This image from 1959 depicts Indio's final submission, "Adventures of Queen Scheherazade."

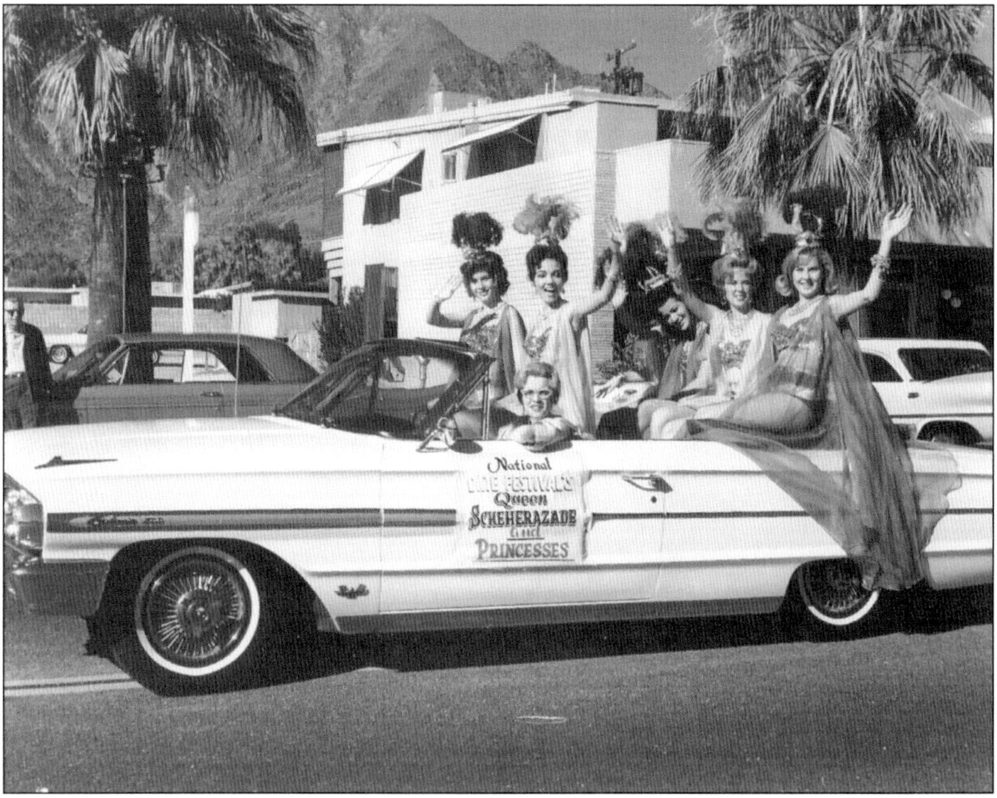

To draw attention to the unique event, Queen Scheherazade contestants occasionally rode in parades elsewhere. Birthed in 1934, the Desert Circus celebrated a Western theme, like early date festivals. The weeklong event culminated in a parade that, given Palm Spring's status as Hollywood's playground, featured celebrities such as Dean Martin and Lucille Ball, and the queen's court. The final Desert Circus was held in 1986.

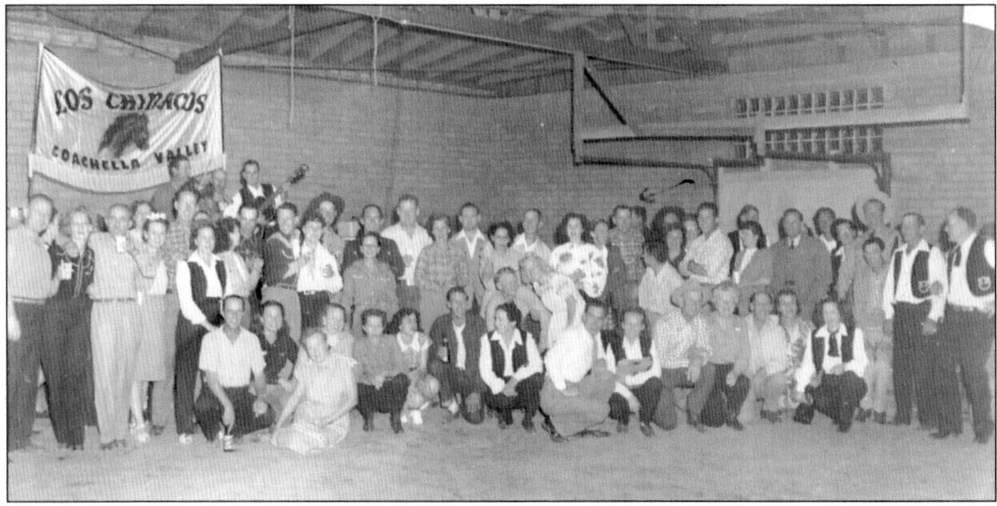

Los Chinacos, Spanish for "outlaws," formed around 1945. The riding club included riders from around the Eastern Coachella Valley. They rode together in the date festival parades, sometimes in Arabian-inspired costume. (Courtesy of the Gilda Guitron family.)

Seven

A Fez for Everyone
Costumes and Community Support

When organizers first proposed an Arabian theme for the date festival, they asked the public to don costumes: harem pants, turbans, and flowing robes. Not only were these outfits less expensive than cowboy gear, boosters argued, they also would draw more attention to the area than Western garb ever could. They were right: the festival's unique costumes worn by Queen Scheherazade and her court, members of the *Arabian Nights* pageant, fair staff, and enthusiastic local residents appeared in regional newspapers, magazines, and, eventually, on television.

Community support was crucial to this publicity push, and it often wowed out-of-town guests. If one had traveled to Indio during fair time in the 1950s, waitresses would likely be wearing harem-girl outfits or bank tellers would be sporting fezzes. Merchants required costumes for their employees both in an attempt to be civic-minded and also in hopes of drawing in more fair-time business. Service clubs and fraternal organizations led the way in terms of costuming, hosting fashion shows, requiring dress for club meetings, and throwing elaborate costumed balls to celebrate the fair. Fabric shops and department stores supported costuming by advertising patterns and offering fabrics inspired by the Middle East. For many years, they even offered in-house costume-making at a discount.

The fair rewarded costumed visitors with free admittance and even had a wardrobe department that clothed the queen, court, pageant participants, fair board, and employees. Outside concessionaries often opted to wear a headpiece or a more elaborate costume. Organizers were careful to point out Hollywood's influence on the costumes and deflect any claims that locals wore authentic Middle Eastern fashion. Some visitors, however, understood these outfits as actual garments worn by those who lived in the Middle East. The costumes also reinforced stereotypes. Most notably the harem-girl outfit, worn by young women at the fair, implied an overt sexuality long associated with the Middle East. By the 1970s, fewer individuals wore costumes at and around the fair. Dressing up continued to fall out of favor as the United States' political and pop-cultural relationship with the Middle East declined at the end of the 20th century.

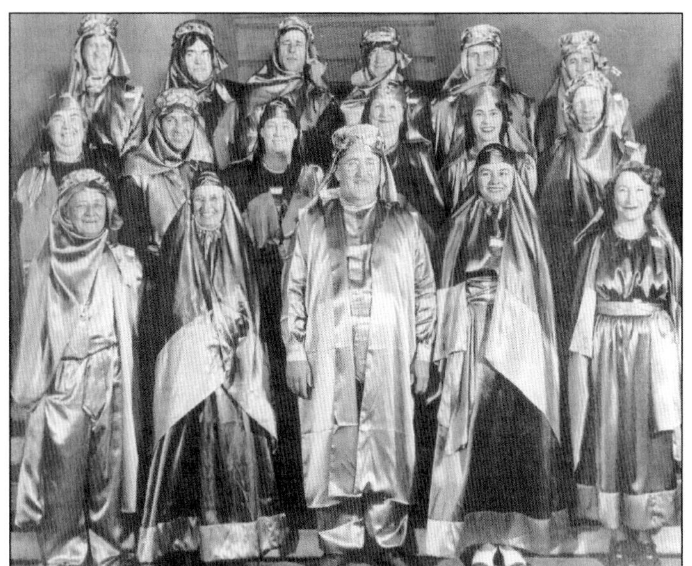

The Indio Civic Club brought the Riverside County Fair to Indio, but members also worked tirelessly to encourage the Arabian theme at postwar date festivals. Here, club members wear satin costumes designed especially for the 1947 fair. Members were business owners and city officials well known in the area who hoped, as in the case of the costumes, to lead by example.

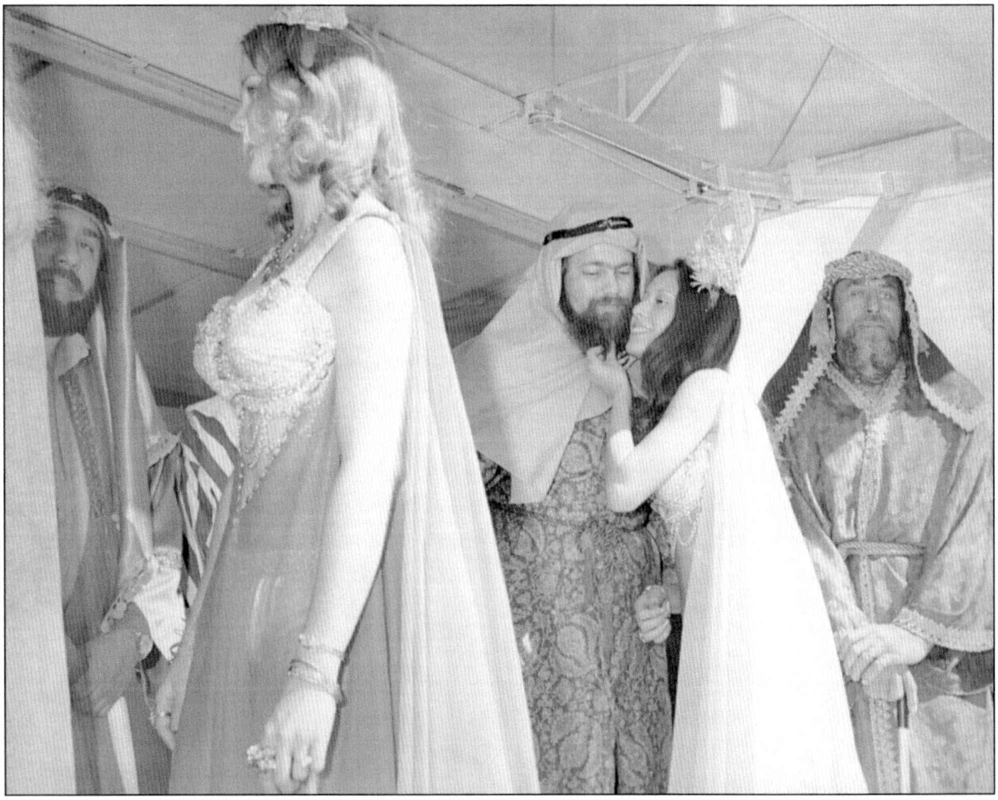

The facial-hair contests continued sporadically after the Western Whiskerino competitions of the late 1930s, morphing into a beardarino challenge. Prizes were awarded for the longest, most colorful, most distinguished, and thickest beards, with competitors donning costumes in an attempt to embrace the Arabian theme. The queen and court served as judges. Dennis Mahr (center) hopes for a win as one of the princesses tests his beard. (Courtesy of the Coachella Valley Water District.)

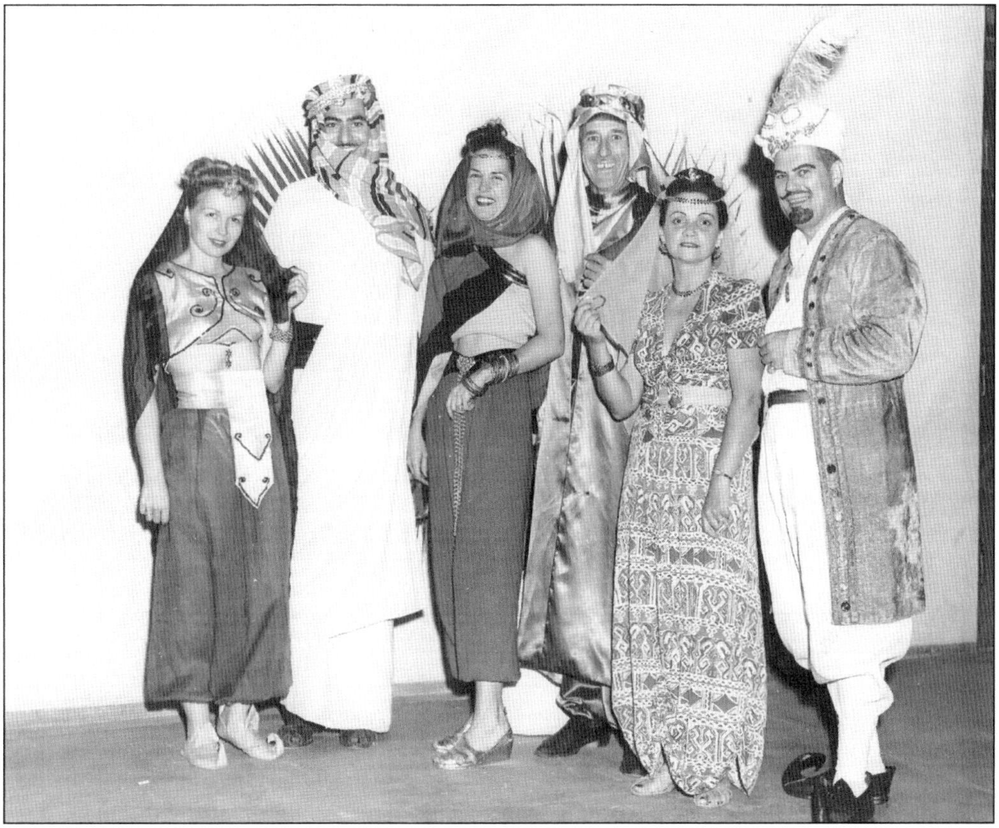

To encourage costuming among residents, locals held prologues to educate the public on what Arabian-inspired costumes might look like. One of these events, held at the Mecca Grammar School in late 1947, featured musical entertainment and a fashion show. More than 200 attended the event. Service and social clubs held fashion shows from the late 1940s into the 1960s, encouraging their members to don the Arabian garments. Pictured above are, from left to right, Margie Feemster, Dr. Edward Albarion, Myrtle Rubidoux, Herbert Ovits, Bessy Ovits, and Dr. C.S. Johnson. Below, children from the school display outfits designed for the younger set.

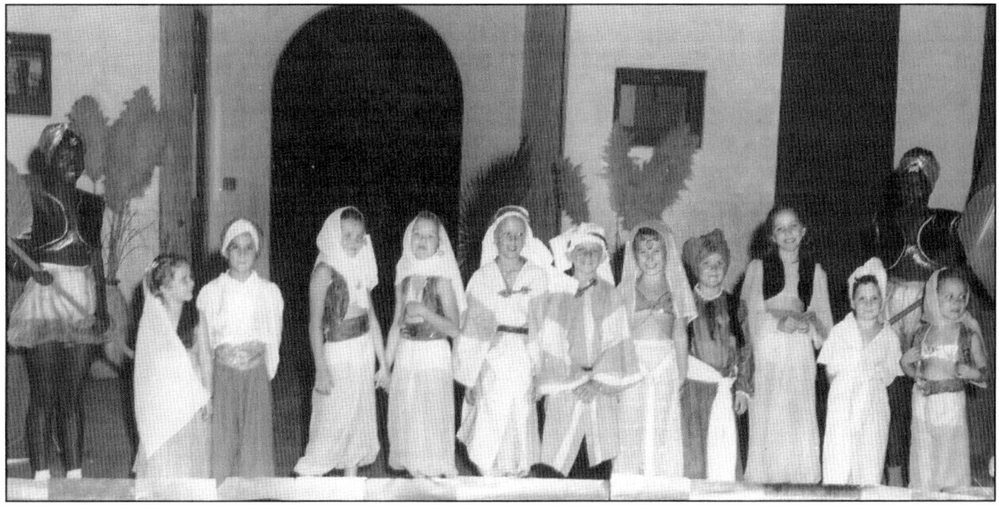

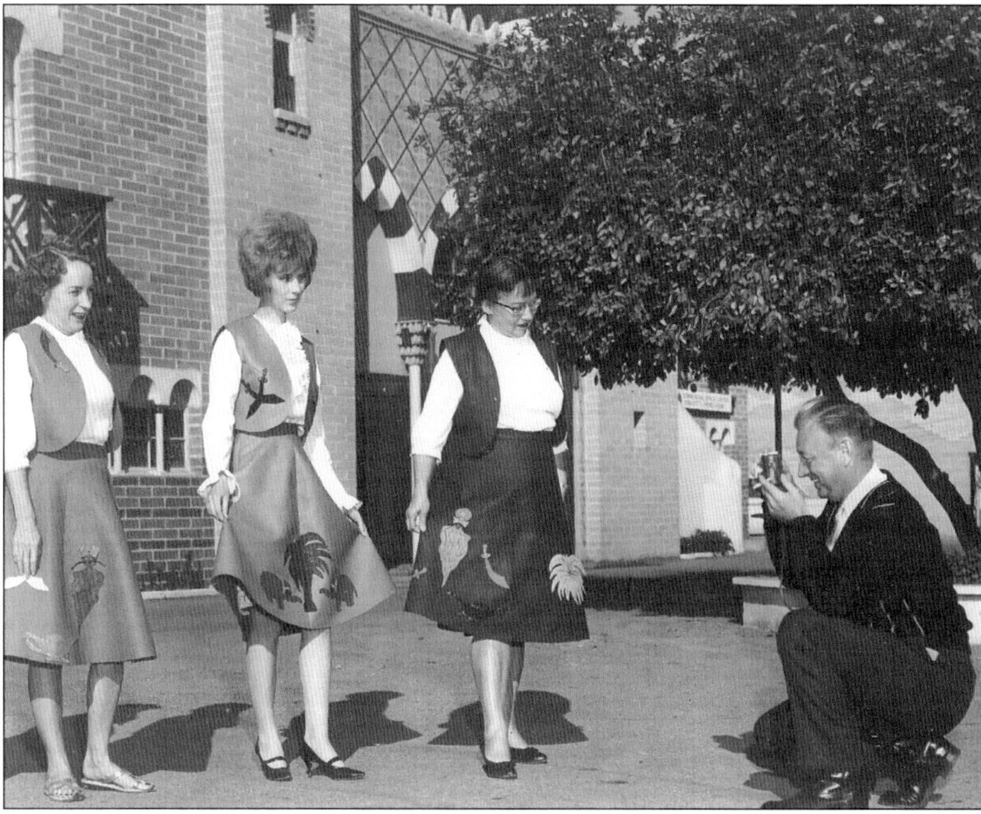

While most costumes mirrored Hollywood imaginings of Middle Eastern wear, some incorporated Arabian inspiration into more traditional American fashion. Here, Charles Wameling, fair manager, photographs concessions staff in front of the Palace of Commerce. The skirts and vests included felt cutouts of palm trees, elephants, scimitars, and a magic lamp and genie.

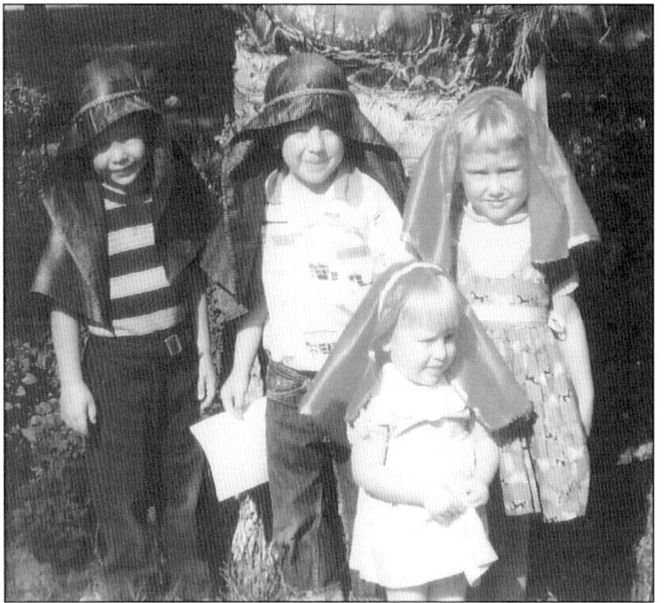

Headpieces were an easy and inexpensive costume piece, often worn on their own. The Anderson children, whose parents owned Colvalda Date Company, pose in their headpieces in 1957. From left to right are Lyle Anderson, Lee "Johny" Anderson III, Leslye Anderson, and Janice Anderson (in front). The youngest residents of Indio put their costumes to use on more than one occasion; several recalled reusing the outfits for Halloween. (Courtesy of the Lee Anderson family.)

A widely circulated mimeographed booklet of costumes provided instructions on how to make the Middle Eastern–inspired outfits. The booklet listed the material length needed and estimated cost of caftans and boleros, while using Middle Eastern names, although misspelled, for other items like the *keffiyeh* (spelled kufigah in the book.) This illustration suggests the addition of trim or a hat could make modern wear into Arabian costumes.

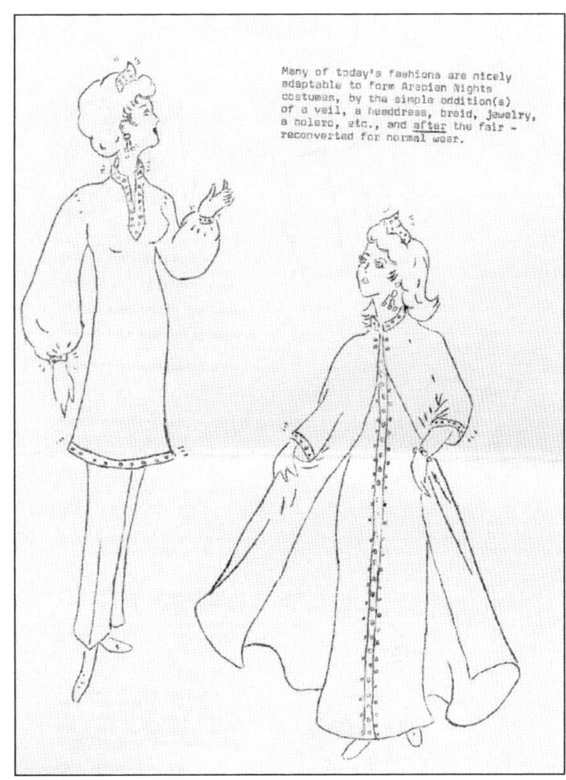

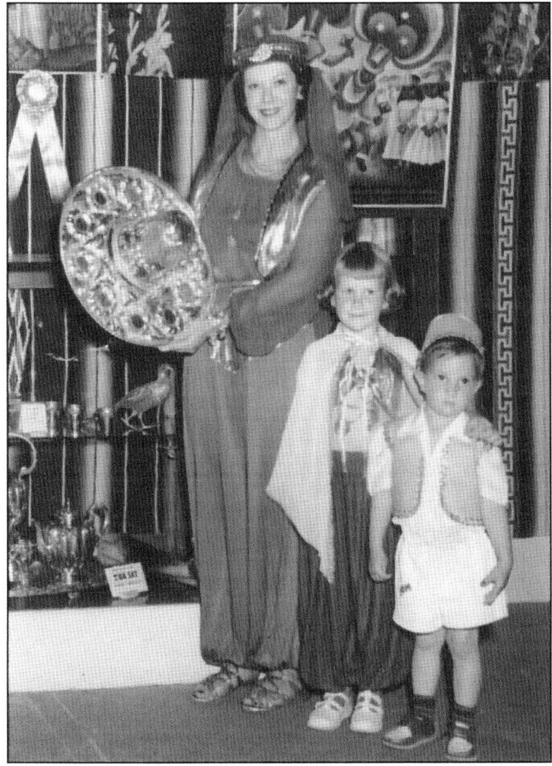

Mothers often crafted outfits for their whole families. Sheldon's, a popular fabric store, advised women on designs and stocked inexpensive fabrics in keeping with the theme. The costumes stand out among the Mexican Coordinating Council's 1950 booth, which featured silverwork from Taxco and Queretaro, Mexico. (Author's collection.)

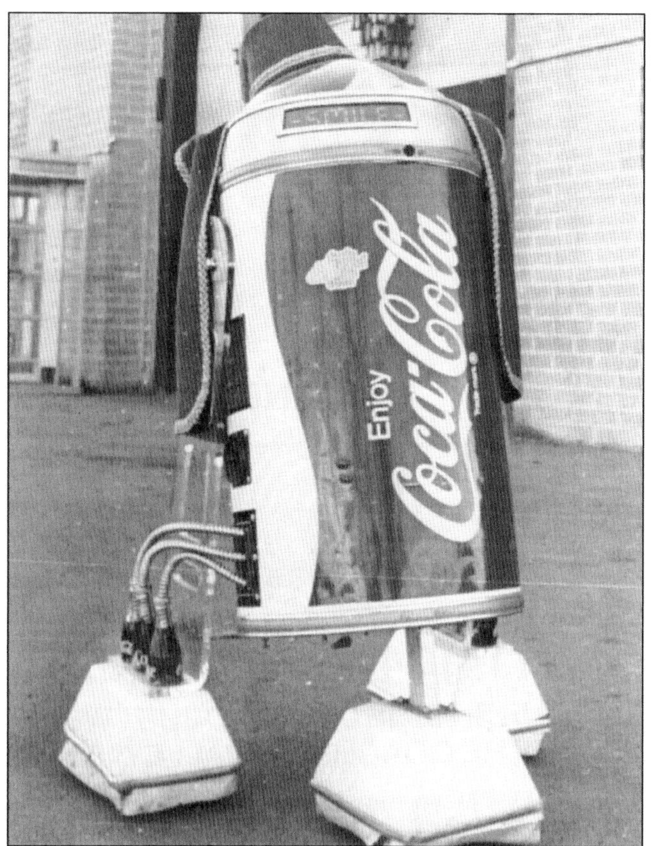

Even the nonhumans got in on the act. The R2D2–inspired Coca-Cola robot dons the signature bolero vest and fez for the festival in the early 1980s at the peak of *Star Wars* fame. The "cobot" played music (via a hidden eight-track player), flashed lights, and moved independently. It made appearances at markets, malls, and events to sell the soft drink. The company offered miniature remote-controlled versions for purchase. (Author's collection.)

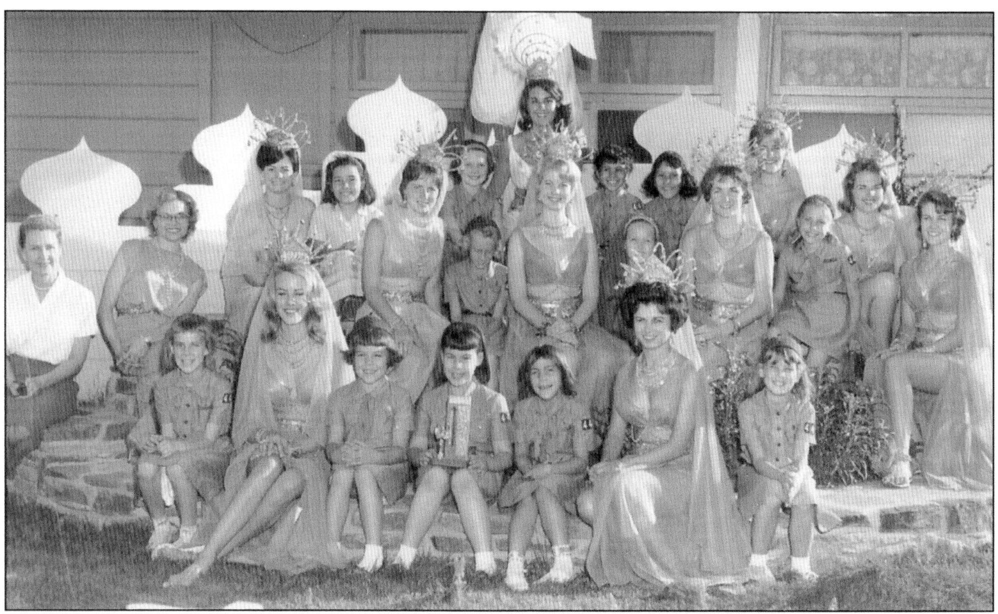

Area Scouts recall obtaining free admission while wearing their uniforms. Here, local Girl Scout Troop 443 are guests of the queen and her court. Scouts also entered exhibits in the fair; a Brownie holds an exhibit trophy in the bottom center. (Photograph by Roy Gillman.)

Employees often wore costumes to work. By 1950, more than 80 percent of fairgoers came from beyond Banning, making the tourist crucial for financial success during the fair. Organizers wanted businesses that interacted with tourists such as service stations, cafés, date shops, and markets to require their workforce to switch their uniforms for Arabian-inspired clothing. In this image, the female staff members at the First National Bank sport their costumes at work.

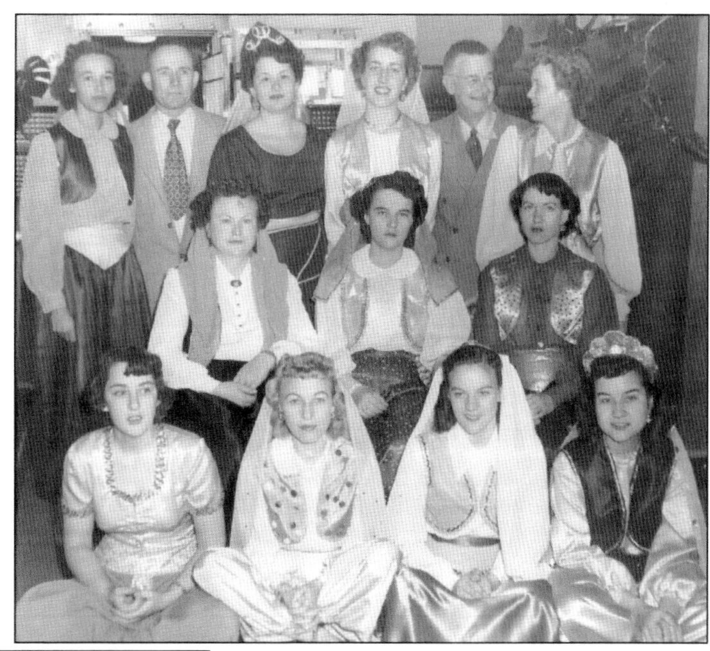

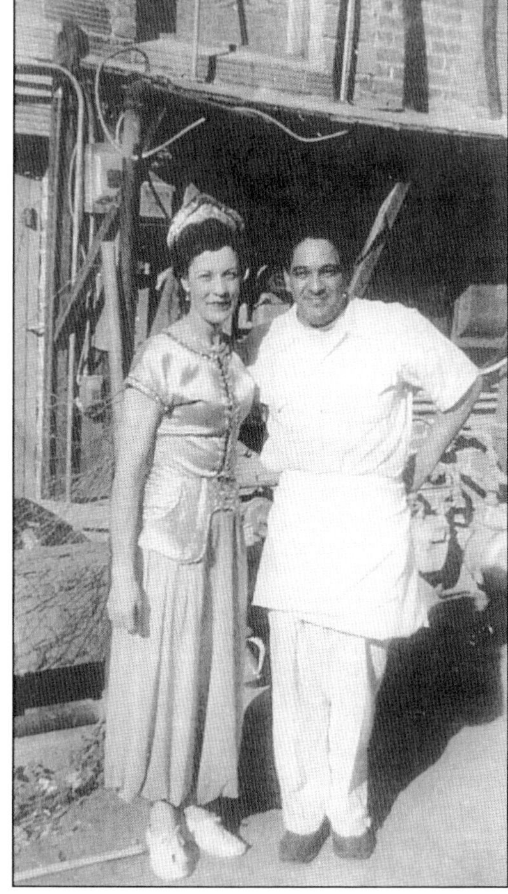

Popular with locals and tourists, Indio's Potter Hotel Coffee Shop drew crowds during the fair. Here, Rudy Villegas III, the fry cook, poses with one of the waitresses in costume, about 1950. (Photograph by Jimmy Smith, courtesy of the Villegas and Gabriel families.)

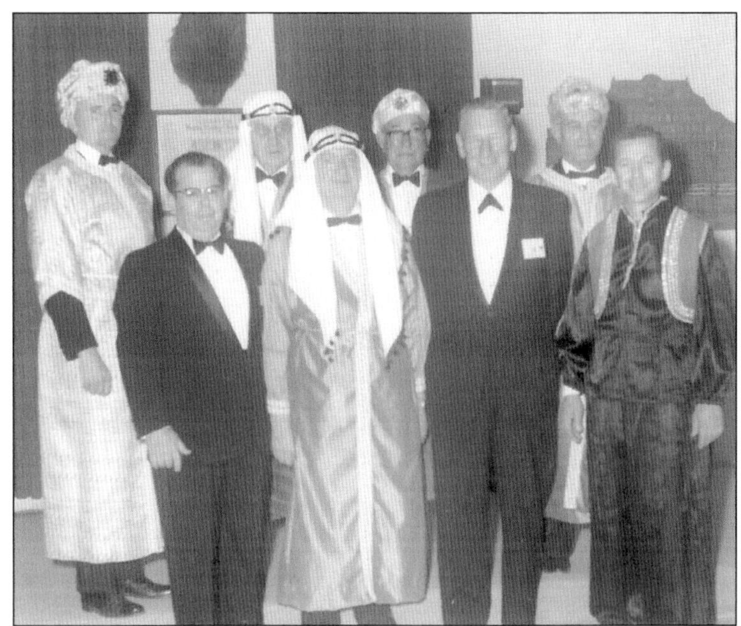

While many fraternal organizations turned to Orientalism for their secret ceremonies and ritual costumes, those located in the Coachella Valley had an additional layer of the Middle East attached. Here, members of the Benevolent and Protective Order of Elks pose in their lodge, some in Arabian costume. (Courtesy of the Asker family.)

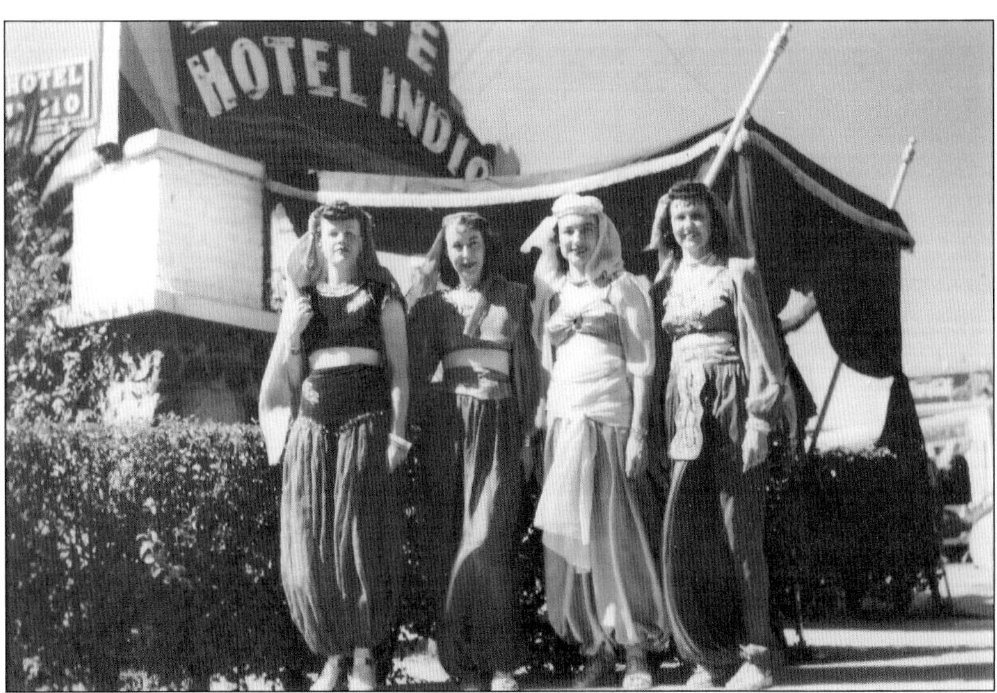

Residents pose in front of the Hotel Indio, with costumes typical of the time. The Hotel Indio served as a home base for Queen Scheherazade and her court during the late 1940s and hosted many out-of-town guests during fair time.

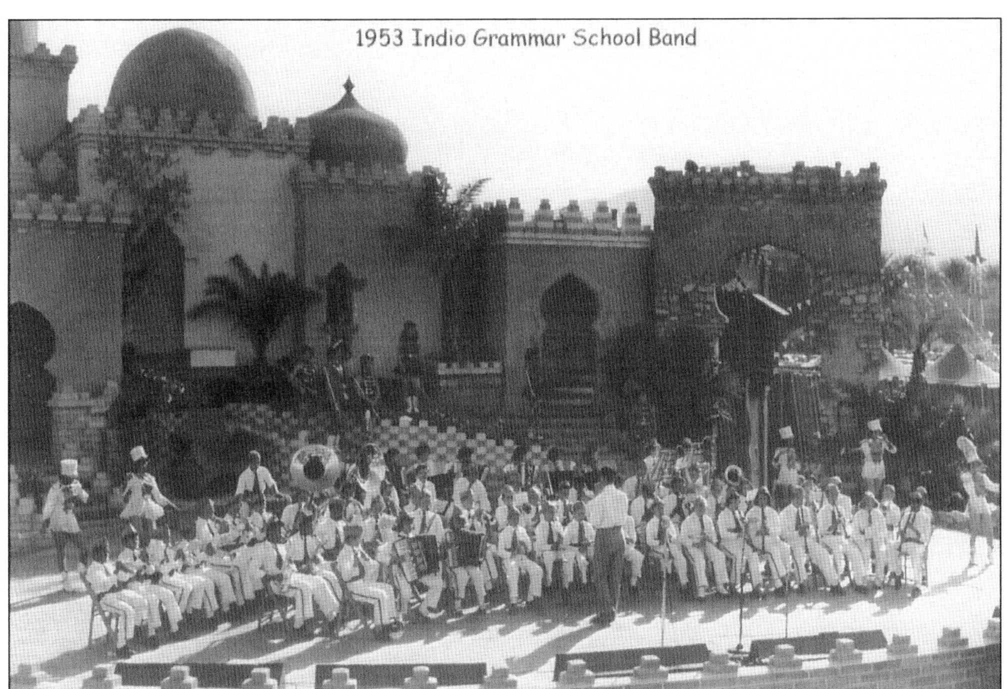

Community participation involved more than costumes. Locals performed and played at the festival. Above, the Indio Grammar School Band, directed by Charles Yates, plays on the Old Baghdad Stage in 1953. Most often, local school bands participated in the parade while a select few gave concerts at the fairground. Below, children perform at the *Buttons and Bows* tap show by Trixie Jarrett's School of Dance in 1960. Children's dance performances were popular at the date festival. Miss Julie's Dance Studio brought their ballerinas and tap dancers to the event from the 1980s through the 2000s. (Above, courtesy of Phillip Nava; below, courtesy of the Asker family.)

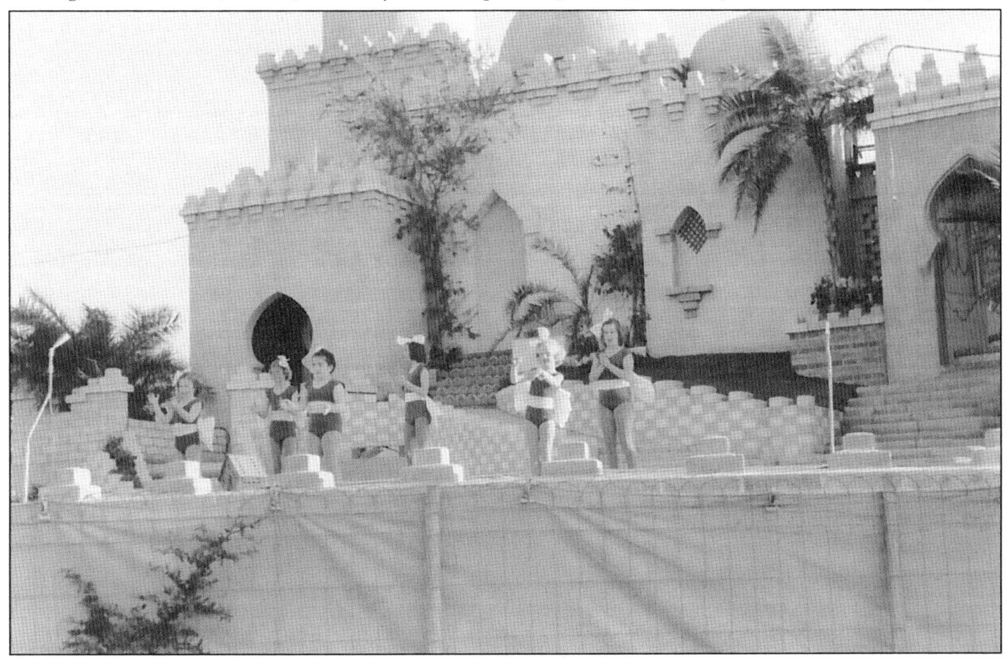

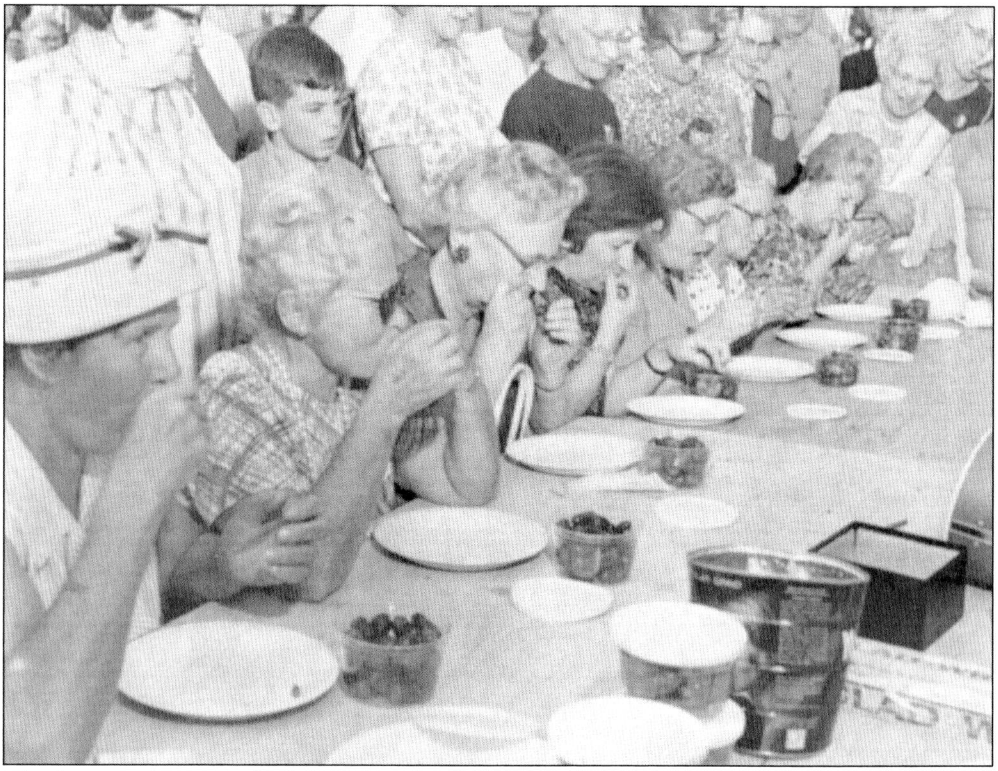

The fair sometimes involved the community with date-related activities. Senior-citizen days, for example, featured date-eating contests. Dates are rich in fiber, potassium, and magnesium, making them an ideal nutritional choice. When planted, a date palm seldom produces the same fruit as its parent. Instead it creates a new variety. One date shop, Shield's Date Gardens, sold over 119 varieties, though just a handful reached the larger commercial market. Deglet Noors, one of the primary varieties grown in the Coachella Valley, were used in the competition. Kids Day offered a date-rolling race, in which children pushed the fruit forward with their noses, pictured here around 1968. (Both, author's collection.)

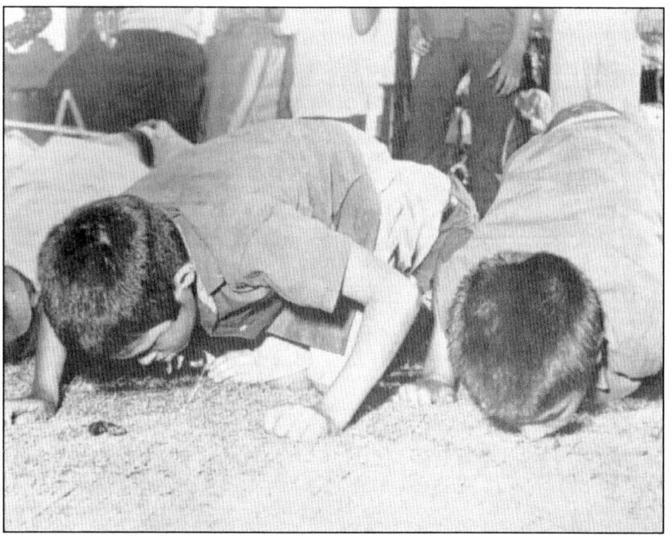

Eight

Desert Dreams
Special Events and People

The date festival sought to celebrate not only the date but the local community as well. Special days and events delighted fairgoers from the start. Salute to Mexico and Latin America days drew interest early on, particularly because of the participation and leadership from the local Mexican and Mexican American community. While the Salute to Mexico events have faded, the festival still courts the large local Latino population with mariachi concerts and targeted vendors. Other special events held during the course of the festival included Canada day, ladies' day, senior-citizen day, photographer day, and armed forces day, some of which offered free or reduced admission to participants. Lifehouse, LeAnn Rimes, Kool and the Gang, Wynonna Judd, and the Commodores performed to blockbuster crowds. The fairgrounds also operated as a venue for several special events that fell outside of fair time. Pow-wows, art festivals, and pioneer picnics sprang to life on-site after the fair had closed. One of Evel Knievel's first public appearances took place in the fairgrounds' large arena in January 1966.

Of course, the date festival hosted its fair share of special guests. Two California governors attended the fair. At least one presidential candidate, Richard Nixon, campaigned at the fairgrounds as well. Politicians who could not attend often declared special date days or weeks, like Gov. Edmund "Pat" Brown's 1959 declaration of "California Date Week," which coincided with the fair. His proclamation mentioned the date festival and encouraged Californians to visit Indio, the Date Capital of the United States. Hollywood stars served as grand marshals for the parades while radio personalities and television hosts shared the charm of the festival with their audiences. Harpo Marx, William Boyd, Leo Carrillo, Duncan Reynaldo, Gary Sinise, Dennis Day, William Devane, and Huell Howser have all been spotted enjoying the delights of the date festival. Even royalty visited: Egyptian prince Abdul Moussa of Khartoum came to the event in 1948 and 1949 while attending the University of Southern California.

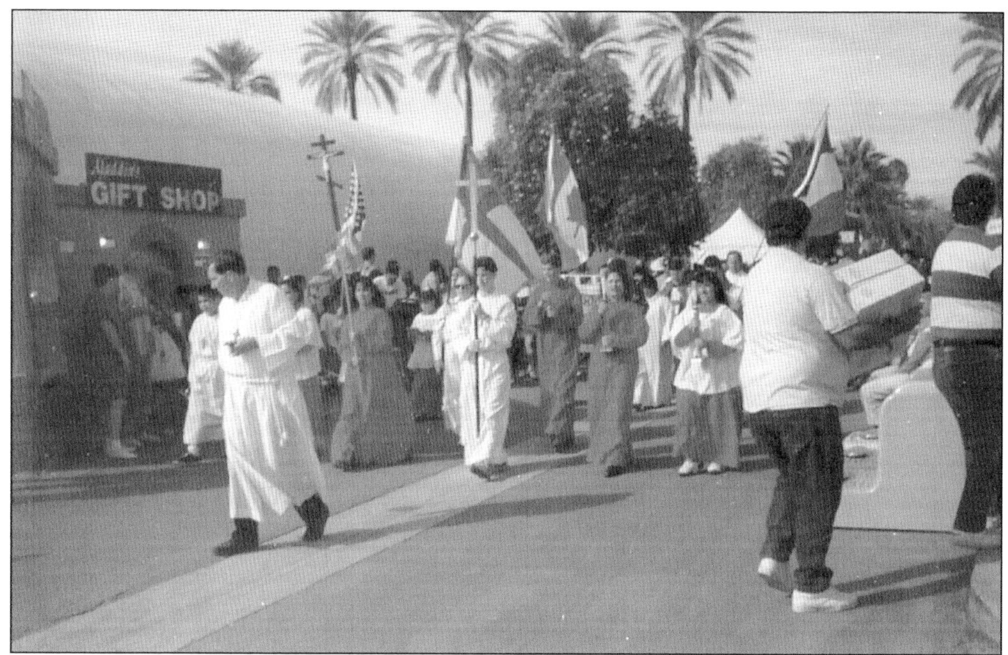

At the Blessing of the Dates, religious figures intoned prayers for the date harvest and the date laborers, known as palmeros. This processional from the mid-1990s arrives at the fairgrounds after beginning at Our Lady of Perpetual Help Parish, roughly a mile away. For a time, the fair also held interdenominational Festival of Faith services on Sunday mornings.

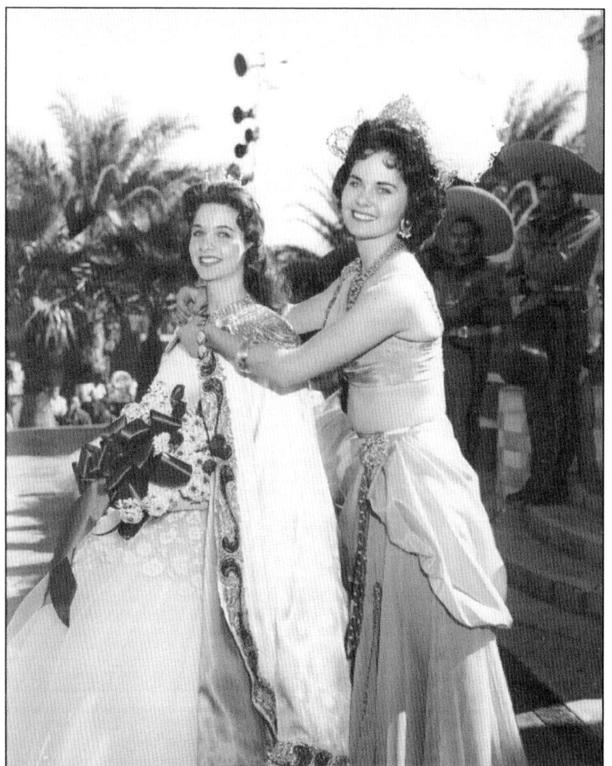

The 1959 date festival hosted a Pan American Celebration on opening day of the fair. Consuls from Mexico, Central America, and South American nations were invited to participate. Karla Sims, of Encino High School, won the title of Pan-American Queen. In this photograph, Queen Scheherazade, Melanie Todd, envelops her in the royal robes. Notice the mariachis in the background. Events celebrating Central and South America often included musical performances. (Photograph by Roy Gillman.)

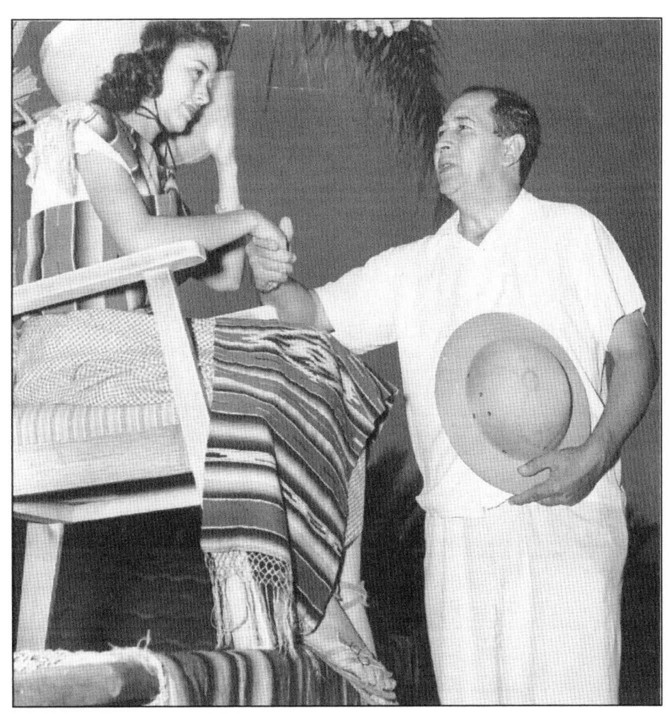

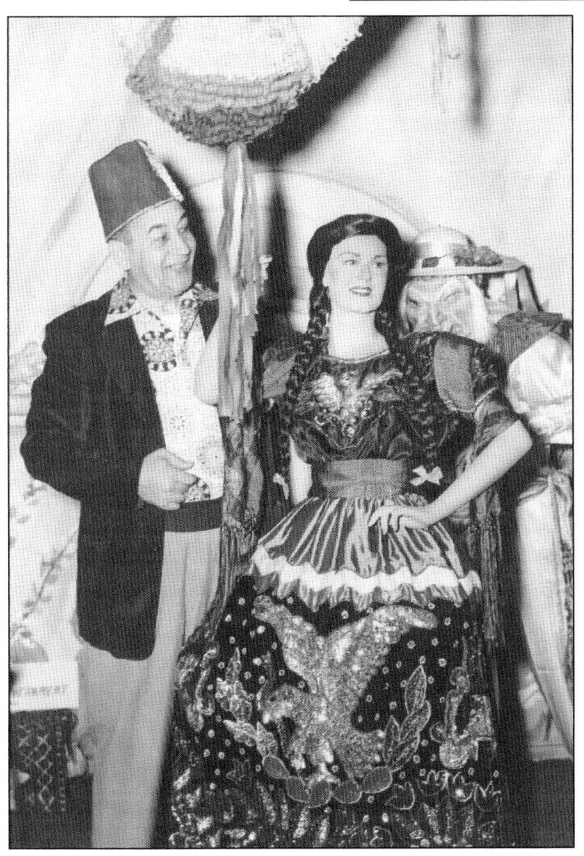

Dr. Reynaldo Carreon organized international events that incorporated Mexico, and their date-growing region in Baja, into the festival. Dr. Carreon, appointed ambassador-at-large to Central and South America by President Eisenhower, worked tirelessly to breed goodwill between the nations. He not only served as acting honorary consul of Mexico in the Coachella Valley beginning in the 1940s but was also awarded Mexico's Order of the Aztec Eagle, the highest honor given to foreigners. Above, Carreon poses with a model during a photo shoot advertising Mexican days at the fair. Below, Carreon oversaw several exhibits designed to educate the public about the history and culture of Mexico. Exhibits paired with special Mexico days, which featured speeches from Mexican governors and other officials, mariachi music, ballet folklorico, and *charreadas*. (Above, courtesy of Linda Beal and the History Hunters; left, courtesy of the Coachella Valley History Museum.)

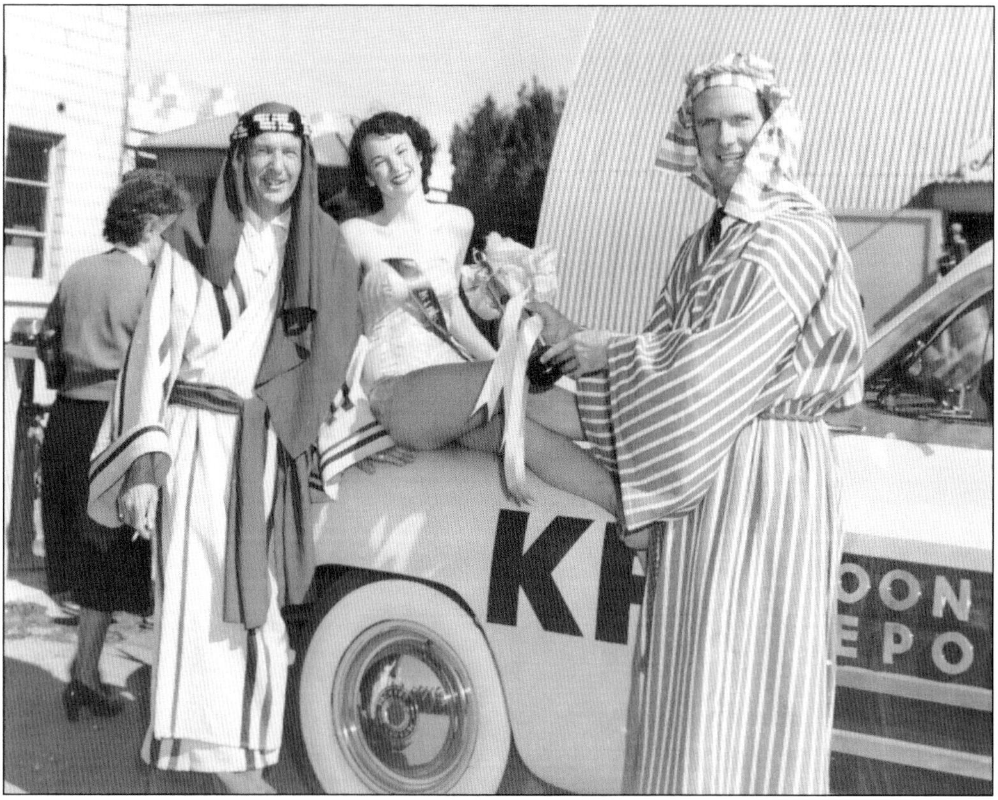

Radio drew in many visitors to the fair. Nelson McIninch featured the agricultural celebration on his *Farm Topics*, *Sunkist Farm Report*, *Noon Farm Reporter*, and *Standard Oil Farm Highlights* for over a decade beginning in 1947. During the live broadcasts, McIninch interviewed fair organizers and young Future Farmers of America. McIninch, in costume, presents an award to 1949's Queen Scheherazade Winnie Gavin, seated next to fair manager Robert Fullenwider. (Author's collection.)

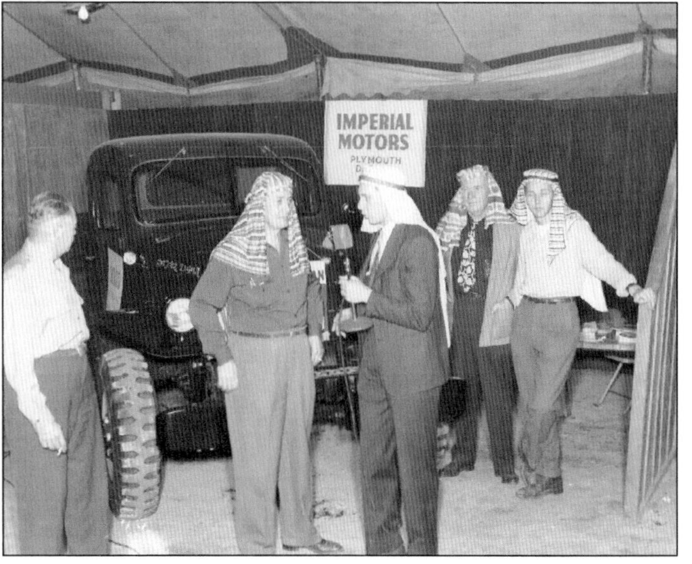

Radio reached potential visitors from around Southern California, though more local radio stations also broadcast from the fair via mobile units. Here, an announcer from KREO radio of Indio broadcasts from the Imperial Motors booth and interviews a car salesman. Given the canvas roofing and strung lighting, the exhibit likely occupied one of the fair's temporary tents, rather than one of the permanent buildings. (Photograph by Bradshaw.)

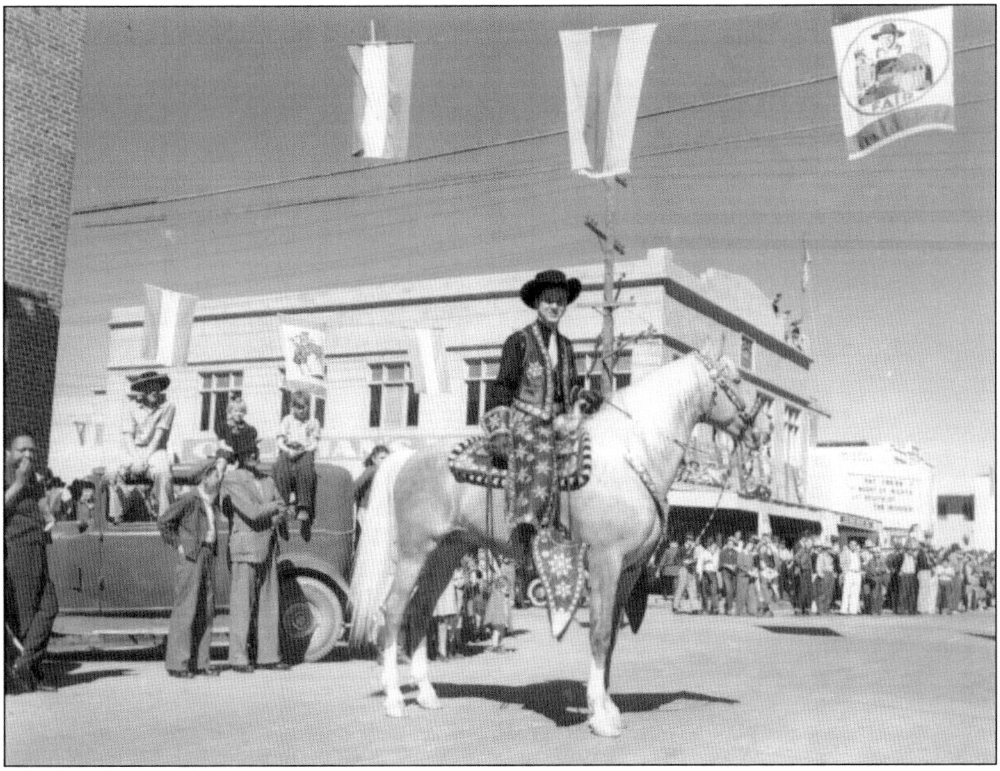

Jackie Cooper served as grand marshal for the 1940 parade, when the fair had a Western theme. Cooper's role in the film *Skippy* (1931) made him the first child actor nominated for an Academy Award, at just nine years old. Well known for his performances in *Our Gang* films, Cooper went on to a successful naval career followed by more acting, producing, and directing in television. (Photograph by Ward's Camera Shop.)

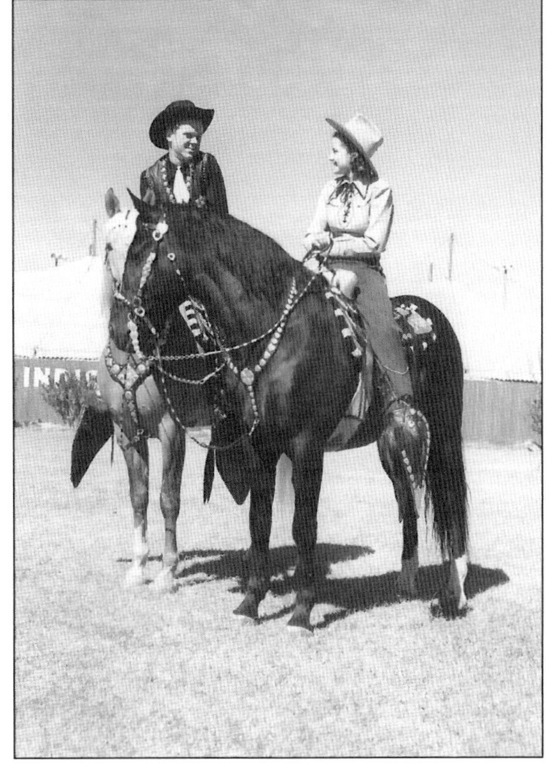

Jackie Cooper brought along his girlfriend Jimmie Rogers for the 1940 parade. Rogers had served as the Palm Springs Rodeo Queen just one month prior. Palm Springs, playground of Hollywood's stars, hosted rodeo and desert circus parades with celebrity grand marshals. Some would participate in Indio's parades as well.

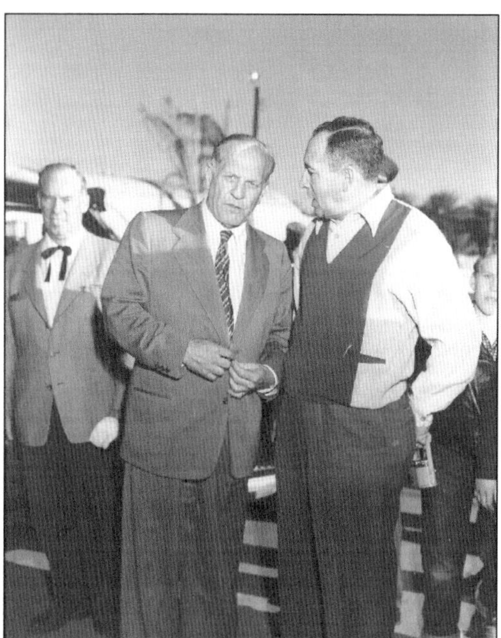

The 31st governor of California, Goodwin Knight, visited the date festival in 1954 at the invitation of the Riverside County Board of Supervisors. Sitting president Dwight Eisenhower was then vacationing in Palm Springs, hence the governor's trip to the region. The media reported that Knight was the first governor to visit the fair, though Gov. Culbert Olson attended in 1940. Here, Knight (left) stands with Dr. Reynaldo Carreon. (Photograph by Roy Gillman.)

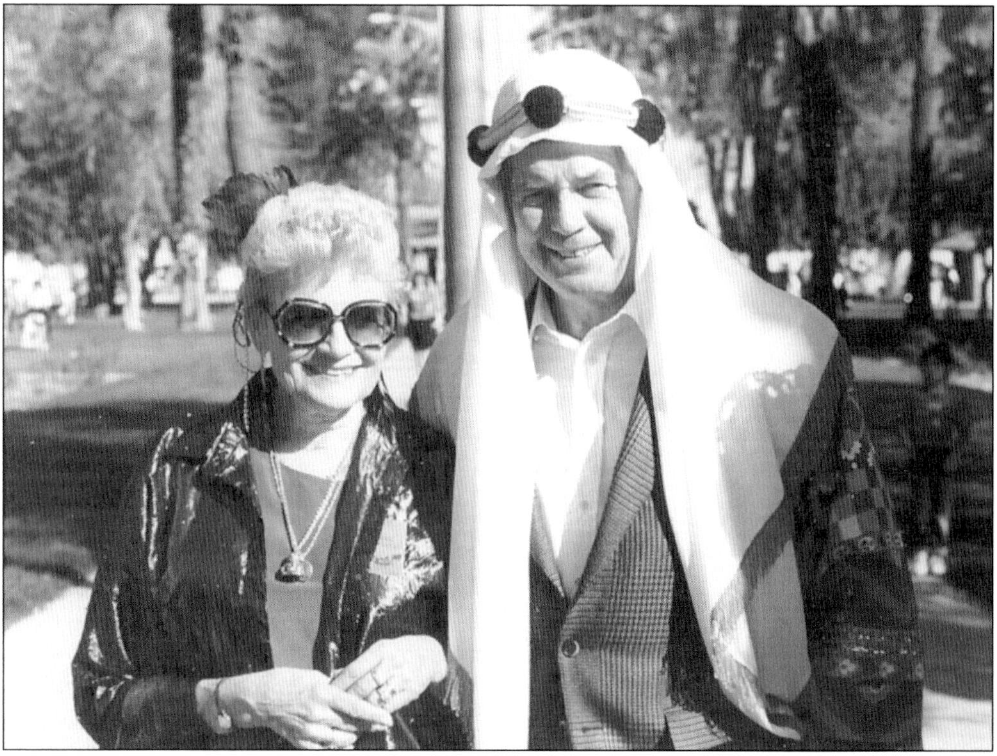

Ralph Waite acted as grand marshal in 1986. Well known for his role as John Walton in *The Waltons*, Waite also ran for political office, including a 1998 losing bid to fill Sonny Bono's congressional seat in the 44th district. More recently remembered for his role on *NCIS*, Waite passed away in Palm Desert in 2014. He is pictured here with Jeri Taylor, fair publicist, in front of Miles Park in Indio.

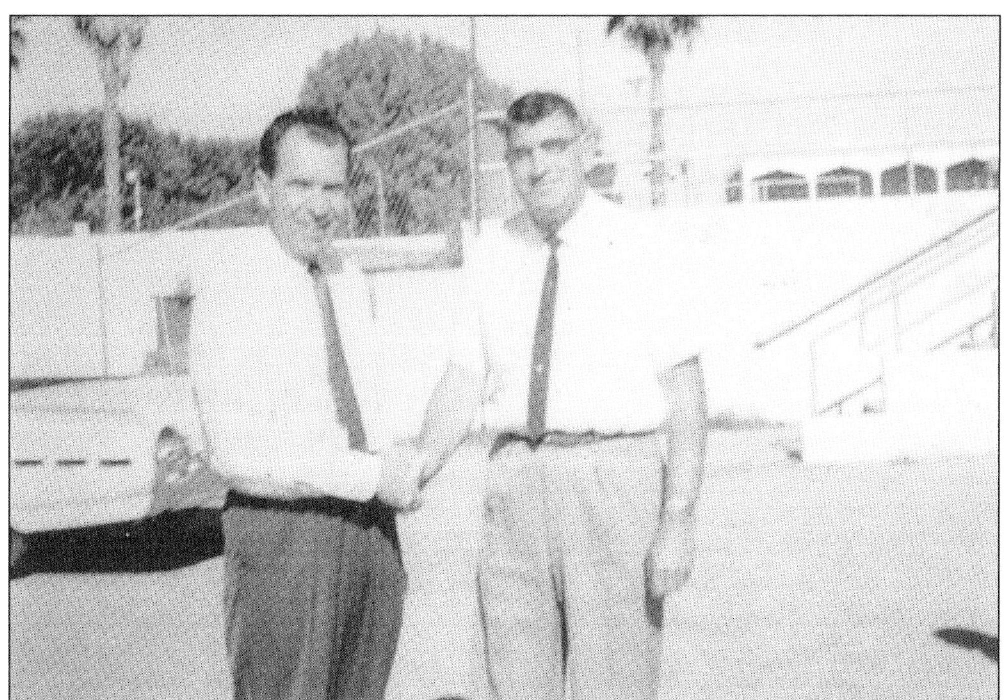

Richard Nixon campaigned in Indio during his 1962 bid for the governorship of California. Though he lost the race to incumbent Edmund "Pat" Brown, he carried Riverside County with nearly 52 percent of the vote. In the top photograph, Nixon poses at the fairgrounds with Elton Gebhardt, his campaign manager for Riverside County. Gebhardt was a rancher in Thermal. In the second photograph, Nixon supporters don festival attire. The date and circumstances of the photograph are unknown. The image may have been taken while Nixon was president or even after, perhaps at his hometowns of Yorba Linda or Whittier, California. (Above, courtesy of Margaret Tyler.)

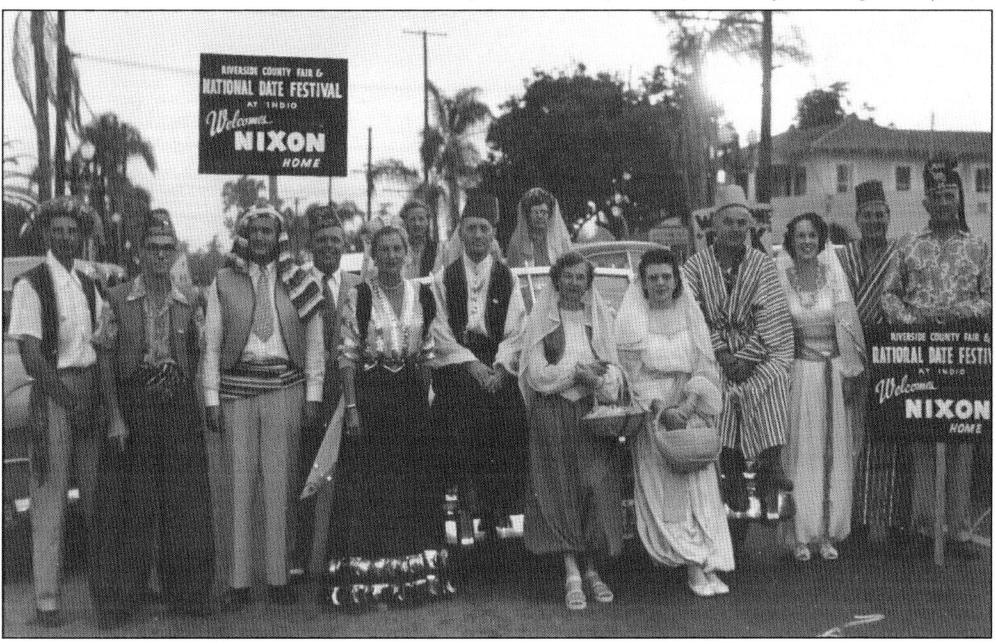

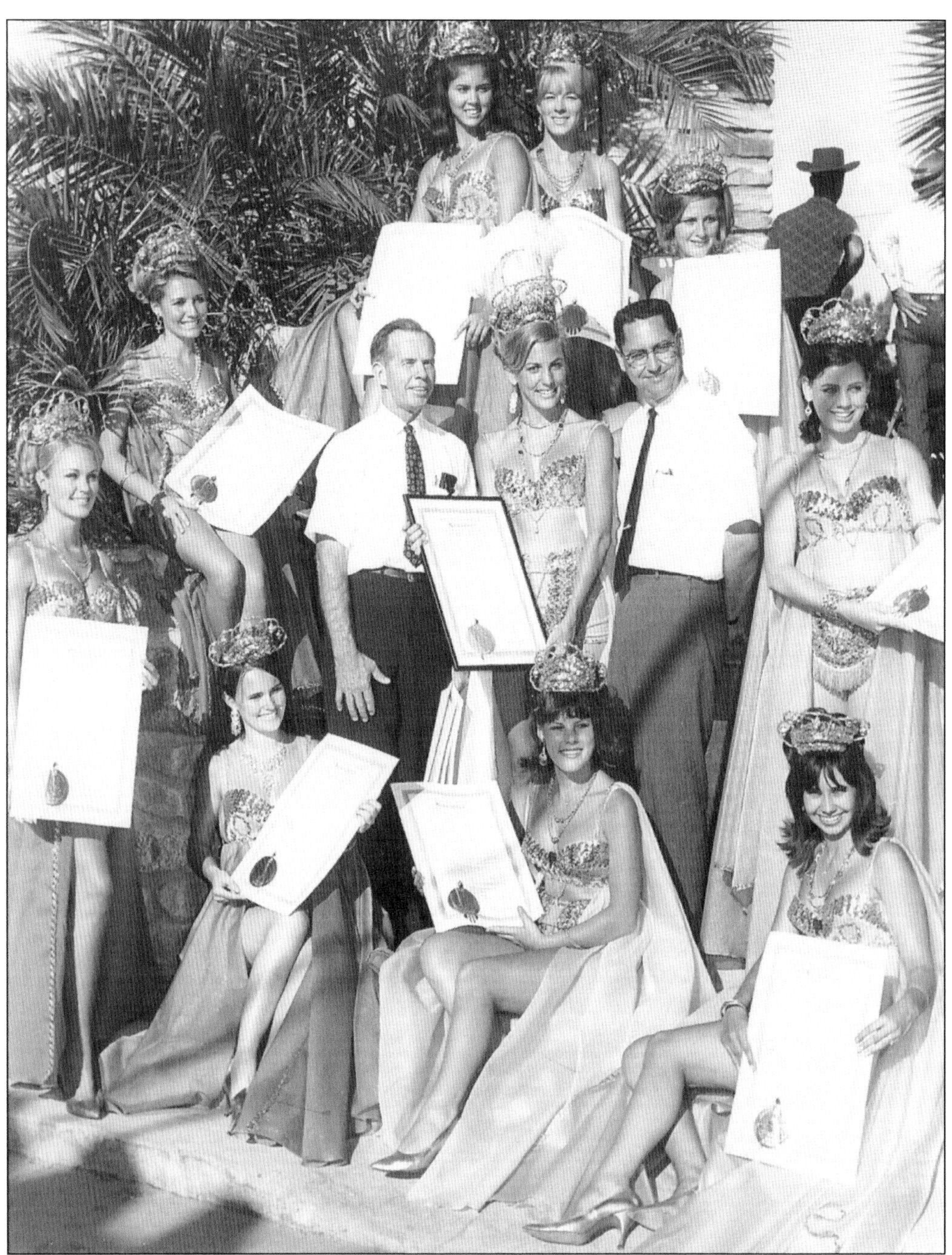

After serving as the mayor of Indio, Gordon Cologne went on to hold office in both the California State Assembly and Senate. A graduate of Coachella Valley High School, Cologne was well known for his commitment to conservation and his work on water resources. His parents, who arrived in the Coachella Valley in 1907 and 1918, owned and operated a date ranch in addition to local meat markets. Here, Queen Scheherazade and her court hold resolutions from the state legislature alongside Cologne, standing right. To the left is Victor Veysey, a California State Assembly member from the Imperial Valley who later went on to serve as a US Congressman in the 1970s.

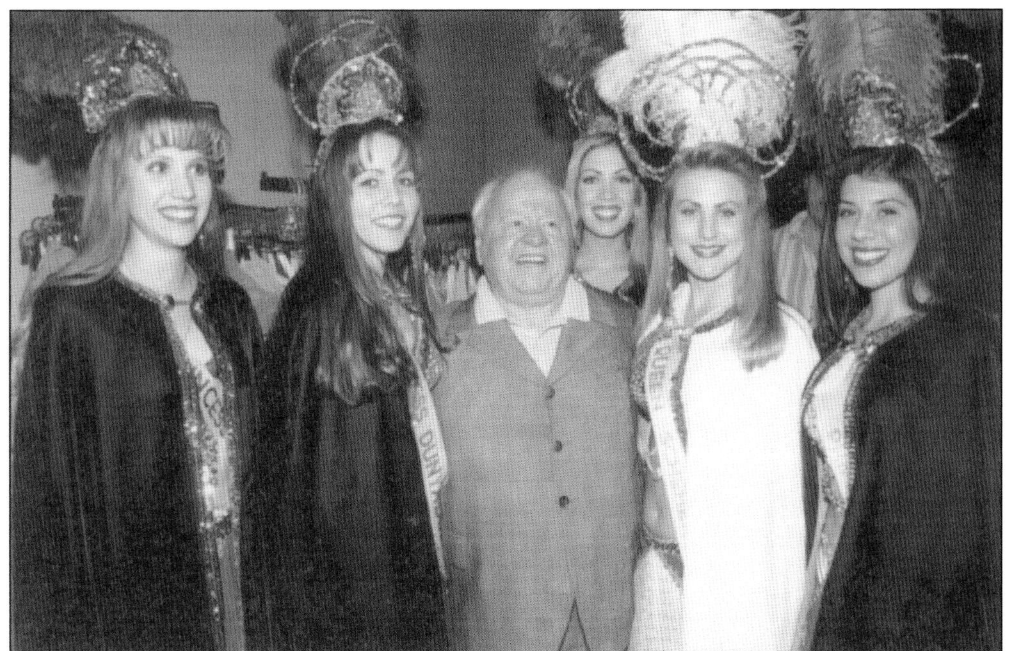

Mickey Rooney visited the date festival during its 50-year celebration. The Academy Award winner known for his Andy Hardy roles maintained one of the longest motion-picture careers, starring in more than 300 films from 1927 until 2014. He also appeared on Broadway and television, earning both a Golden Globe and Emmy Award. (Courtesy of the Riverside County Fair and National Date Festival.)

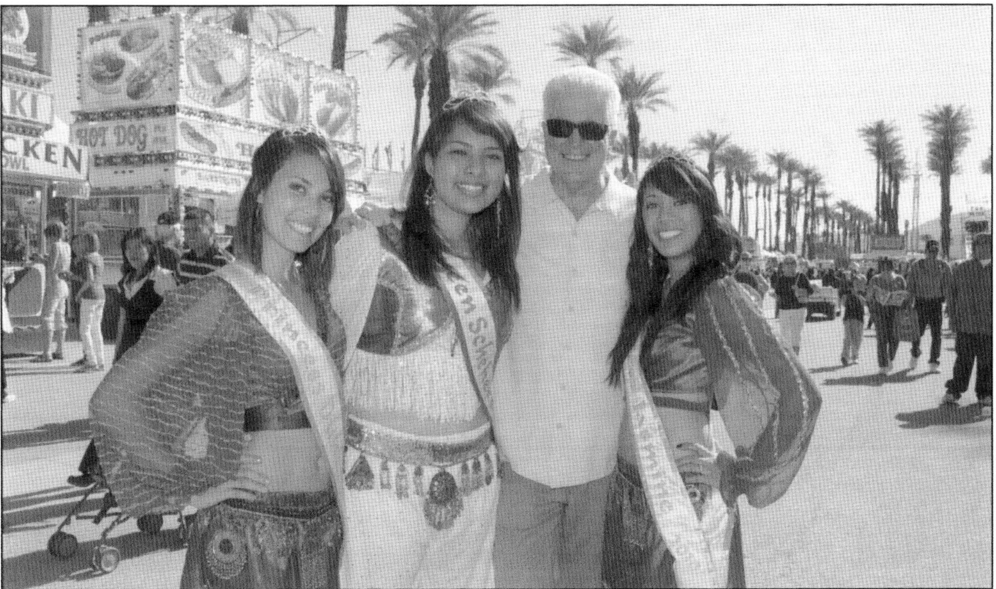

Huell Howser, California's golden boy, filmed an episode of *California's Golden Fairs*, a spinoff of *California's Gold*, at the date festival in 2010. In addition to meeting with the queen and court, Howser spoke with costumed volunteers in the fine-arts building, tasted bites from the annual amateur date recipe contest, enjoyed an "amazing" date shake, and hobnobbed with camels. (Courtesy of the Riverside County Fair and National Date Festival.)

Bibliography

Coachella Valley News and Indio Index. Indio, CA.
Coachella Valley Submarine. Indio, CA.
Date History Collection. Coachella Valley History Museum Archives.
Desert Sun. Palm Springs, CA.
Ephemera and Vertical File Collection. Indio Public Library.
Ephemera and Vertical File Collection. Riverside Public Library.
Indio Daily News. Indio, CA.
Indio News. Indio, CA.
Laflin, Patricia. "The Story of Dates Part I and II." *Periscope* (2006, 2007).
Los Angeles Times. Los Angeles, CA.
Press Enterprise. Riverside, CA.
Seekatz, Sarah. "America's Arabia: The Date Industry and the Cultivation of Middle Eastern Fantasies in the Deserts of Southern California." PhD. diss., University of California, Riverside, 2014.

About the Coachella Valley History Museum

Founded by the Coachella Valley Historical Society in 1984, the Coachella Valley History Museum seeks to preserve, interpret, and share the rich history of Southern California's Coachella Valley. The museum's Indio campus includes a thriving archival complex, antique farm equipment, the historic Smiley-Tyler adobe, Indio's 1909 schoolhouse, a blacksmith shop, a desert submarine, extensive gardens, and the nation's only date museum. Primarily a volunteer-run nonprofit, the campus also hosts exhibits about the Eastern Coachella Valley's Native American populations, the region's relationship to precious water resources, its early economic development via agriculture (especially the date) and tourism, the development of regional infrastructure and railroads, and even a home kitchen set in the 1930s. Additionally, the Coachella Valley History Museum supports diverse groups and events, including Mexican American pioneer oral histories and exhibits, the Black Pioneer Project, the Junior Historians Program, school tours, art exhibits, living history events, and much more.

Currently, the Coachella Valley History Museum is open to the public Thursday through Saturday from 10:00 a.m. to 4:00 p.m. and from 1:00 p.m. to 4:00 p.m. on Sundays, from October through May, though it is available at other times for school and group tours, archival research, and events. Learn more about the museum at www.cvhm.org or by calling (760) 342-6651.

Discover Thousands of Local History Books Featuring Millions of Vintage Images

Arcadia Publishing, the leading local history publisher in the United States, is committed to making history accessible and meaningful through publishing books that celebrate and preserve the heritage of America's people and places.

Find more books like this at
www.arcadiapublishing.com

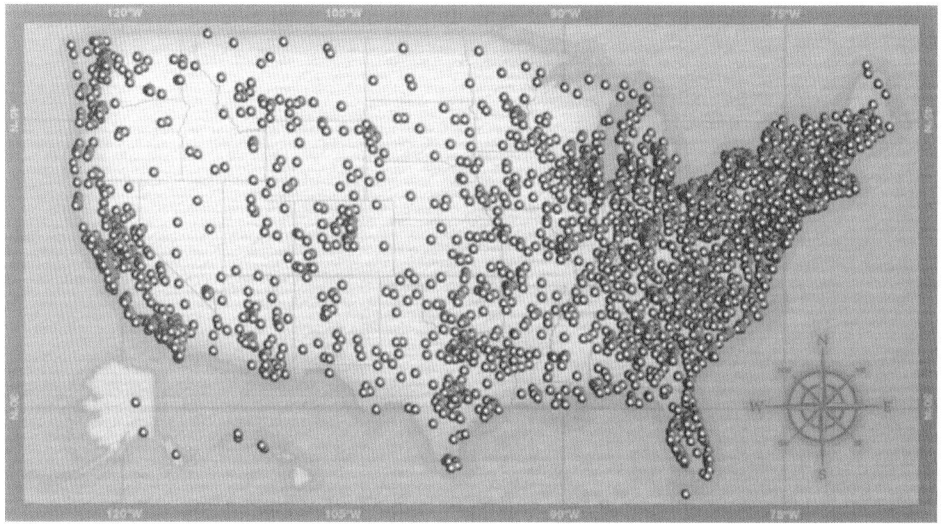

Search for your hometown history, your old stomping grounds, and even your favorite sports team.

Consistent with our mission to preserve history on a local level, this book was printed in South Carolina on American-made paper and manufactured entirely in the United States. Products carrying the accredited Forest Stewardship Council (FSC) label are printed on 100 percent FSC-certified paper.

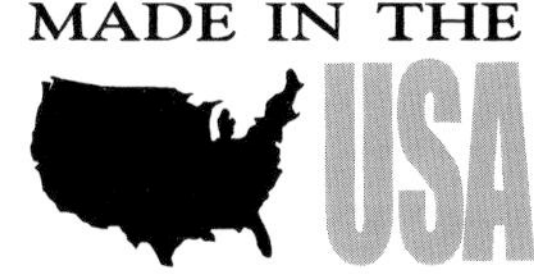